This Artful Sport

Fly-Fishing Books by Paul Schullery

Nonfiction

American Fly Fishing: A History
Royal Coachman: The Lore and Legends of Fly Fishing
Cowboy Trout: Western Fly Fishing As If It Matters
The Rise: Streamside Observations on Trout, Flies, and Fly Fishing
If Fish Could Scream: An Angler's Search for the Future of Fly Fishing
Fly-Fishing Secrets of the Ancients: A Celebration of Five Centuries of Angling Lore and Wisdom
The Fishing Life: An Angler's Tales of Wild Rivers and Other Restless Metaphors

Fiction

Shupton's Fancy: A Tale of the Fly-Fishing Obsession
A Fish Come True: Fables, Farces, and Fantasies for the Hopeful Angler

Fly-Fishing Books by Steve Raymond

Nonfiction

Kamloops: An Angler's Study of the Kamloops Trout
The Year of the Angler
The Year of the Trout
Backcasts: A History of the Washington Fly Fishing Club, 1939–1989
Steelhead Country
The Estuary Flyfisher
Rivers of the Heart
Blue Upright
Nervous Water: Variations on a Theme of Fly Fishing
A Fly Fisher's Sixty Seasons

Fiction

Trout Quintet: Five Stories of Life, Liberty and the Pursuit of Fly Fishing
Six Fish Limit: Stories from the Far Side of Fly Fishing

This Artful Sport

A GUIDE TO WRITING ABOUT FLY FISHING

Paul Schullery
and Steve Raymond

Essex, Connecticut

An imprint of The Globe Pequot Publishing Group, Inc.
64 South Main Street
Essex, CT 06426
www.globepequot.com

Distributed by NATIONAL BOOK NETWORK

British Library Cataloguing in Publication Information Available

Library of Congress Cataloging-in-Publication Data Available
ISBN 9781493085378 (paperback) | ISBN 9781493085385 (epub)

∞™ The paper used in this publication meets the minimum requirements of American National Standard for Information Sciences—Permanence of Paper for Printed Library Materials, ANSI/NISO Z39.48-1992.

For Nick Lyons: Author, editor, publisher,
and a dear friend and inspiration to both of us.

Contents

Books are certainly one of the most dependable sources of companionship an angler can have.

—Arnold Gingrich, *The Joys of Trout*

First Words

The first printed English text about fly fishing, long attributed to Dame Juliana Berners, called it "this artful sport." It has never been firmly established whether Juliana Berners really existed, or if she (or he?) might have been one of several authors of the seminal *Treatyse of Fishing wyth an Angle*, but there can be no doubt the description was accurate: Fly fishing *is* an artful and art-filled sport, having inspired painting, photography, sculpture, poetry, woodwork, writing, and an enduring treasury of books.

Today fly fishing also inspires a daily freshet of magazine articles, newsletters, internet blogs, online forums, podcasts, and a continual flow of news, comment and debate, recording the culture, history, tradition and technology of the sport. Nevertheless, you can diligently search all these sources and find few, if any, that explain the mysterious quality of fly fishing that inspires so many people to write about it.

What *is* behind this mysterious quality? Perhaps it has to do with the rewards of fly fishing, which are many, well known and may offer a more profound sense of self-fulfillment than any other sport. Some anglers grow to love fly fishing so much it makes them feel what amounts almost to a moral obligation to share their feelings, observations and discoveries, even with people they've never met—an impulse not to be denied. So although you can find books or articles about almost every other aspect of the sport—casting, fly tying, history, entomology, personal experience, philosophy, technology, technique, individual waters, fish species, ecology, or just about anything else you want to know—you'll find very few words, if any, that explain the motivations of those who write about it, or how they do it.

Until now.

This book is a guide for those who wish to write about this revered sport—or perhaps, in some cases, why they shouldn't. It tells how to write for print or online magazines and traditional or self-published books. It describes the equipment you'll need, how to develop your writing "voice," choose a publisher and market your work. It explains how to select an agent if you need one, what to do to promote your published work, and how to deal with reviews and reviewers. It will even help you minimize the number of wind knots in your fly-fishing prose.

But while offering as much help as we can, there are quite a number of wheels we do not intend to reinvent here. There is a tremendous amount of helpful advice and information for writers already out there in books, websites, and other resources that, although they say little or nothing specifically about fly-fishing writing, are still invaluable for writers in general. We're focusing on fly-fishing writing in this book, but whenever it seems appropriate, we will direct you to such resources.

Who are we to offer such advice? Well, we've both been there and done that and have the scars to prove it. Paul Schullery, author, coauthor, or editor of more than 50 books, is former executive director of the American Museum of Fly Fishing and editor of its journal, *The American Fly Fisher.* His nonfiction fly-fishing books include *American Fly Fishing: A History*, *Royal Coachman*, *Cowboy Trout*, *The Rise*, *If Fish Could Scream*, and *Fly-Fishing Secrets of the*

Ancients. His fictional fly-fishing works include *Shupton's Fancy* and *A Fish Come True*. He wrote and narrated the PBS film *Yellowstone: America's Sacred Wilderness* and was an advisor and interviewee for the Ken Burns film *The National Parks: America's Best Idea*.

Steve Raymond, longtime editor and manager at *The Seattle Times*, also edited two fly-fishing magazines, *The Flyfisher* and *Fly Fishing in Salt Waters*. A reviewer of fly-fishing books for several publications over many years, his work has been published in nine anthologies and at least two dozen magazines, including *Sports Illustrated*, *Fly Fisherman*, and *Western Fly Fishing*. He is author of ten nonfiction books about fly fishing, including *The Year of the Angler*, *The Year of the Trout*, *Steelhead Country*, *The Estuary Fly Fisher*, *Rivers of the Heart*, and *Blue Upright*, plus two collections of fly-fishing fiction, *Trout Quintet* and *Six Fish Limit*.

Between us we've been exposed to virtually all aspects of the publishing business, both as writers and editors. Having learned its lessons, pitfalls and secrets the hard way—by discovering them for ourselves—we hope through this book to make it easier for others to fulfill their fly-fishing publishing aspirations. After all, just because countless fly-fishing articles and books are already in print, it doesn't mean there isn't room for more.

Including yours.

Words define the essence of the sport.
—Steve Raymond, *Rivers of the Heart*

CHAPTER ONE

Getting It Write

Fly fishing and writing are very different disciplines but they enjoy at least two remarkable similarities. Each is a solitary pursuit—when you're alone on the water, it's just you and the fish, and when you sit down at the writing table, it's just you with all those empty pages. Each also relies heavily on individual experience, skill, research and introspection, all as necessary for reading the water, choosing the right fly or making an accurate cast as they are for writing crisp, coherent, entertaining prose for magazine articles or books. We could even add a third similarity: both fly fishing and writing involve a great deal of hope.

You've probably spent a long time learning to become a competent fly fisher, practicing casting and fly tying, absorbing the sport's literature and history, assimilating its nomenclature and ethics, learning your way around rivers, lakes or salt water, and mastering the many little things that make a successful fly angler. You also probably enjoyed the experience, because learning is one of the great rewards of fly fishing, and it never ends.

What you may not realize, however, is that becoming a writer requires a similar learning experience. Many people assume they have natural writing talent only to discover they were mistaken when they try it; even those who *do* have natural talent inevitably face a lengthy apprenticeship to refine and develop it. This includes reading the work of other writers—it's impossible to write well without reading the work of others who have written well—and spending many hours practicing and polishing your prose until you're confident it's as good as can be. It also means subjecting your writing to the constructive criticism of others, then, when you're ready, taking the final step of subjecting your work to the scrutiny (and mercy, if any) of a professional magazine or book editor. All this is a necessary part of developing your own writing "voice," which is vital if you plan to write books and nearly as crucial if you're just aiming to write magazine articles.

What do we mean by "voice?" It's all the elements that combine to form your own unique literary style—simplicity, rhythm, cadence, lyricism, word choice, alliteration, punctuation, nuance and color. These blend seamlessly to make your words as unique as the works of a classical composer, or as distinctive as a songbird's signature call. The best writers—from Hemingway to Dr. Seuss to Roderick Haig-Brown—are instantly identifiable by their prose. That should also be your goal.

Many if not most newcomers find that learning to be a good writer isn't as much fun as learning to be a fly fisher (Paul found both to be simultaneously exasperating and exciting), but this book will try to make it easier and more enjoyable than it would be otherwise; we'll even suggest a few exercises and other little schemes that might help speed up the process.

Make no mistake, though, writing is a tough business, and not everyone has what it takes, including the ambition or stamina, to succeed. Let's find out if you're among those who do.

Why Write: Some Good (or Great) Reasons

We'll start with your motives. As we've seen, you might have a deep-seated desire to share your fly-fishing experiences, descriptions or inspirations with others, especially those just becoming acquainted with the sport. There are hardly any better motives, but there are many others, both good and bad, that cause people to try writing about fly fishing.

Here are some good ones:

You have good news to report, for example the removal of fish-blocking dams in various sections of the country, successful efforts to restore damaged fishery stocks, or some exciting progress made in restoring your local trout stream.

Calling attention to some truly significant new development in fly fishing is always a legitimate reason for writing. Such developments are rare, but they do occur from time to time. The growing popularity of Spey rods in recent years might be one example; developments in technology—GPS systems, electronic hatch charts, and so on—could be another.

Fun is always a righteous motivation. If you have a good, fun subject, you'll enjoy writing about it and can find a way to communicate your enjoyment to readers, don't hesitate to go for it.

Another legitimate reason for writing is to call the public's attention to something that poses a threat to the sport, which usually means an environmental problem. Sometimes these stories turn out well and make a difference, and sometimes they don't. For example, many years ago Steve wrote a story about a massive development proposed for the north shore of Pass Lake, a popular, scenic lake on Fidalgo Island in northern Puget Sound. One of Washington state's first fly-fishing-only waters, Pass Lake was an extremely rich lake that routinely produced trout weighing up to four pounds.

The lake's southern half was protected within the bounds of Deception Pass State Park, but the northern half and its scenic surrounding forests and pastures was owned by a retired couple living on a fixed income. Unable to pay rising property taxes, they began searching for a way to preserve the property, but when that failed they were forced to option it to a developer. The developer announced plans to build town houses all around the north shore along with a supermarket and even a bowling alley. Steve heard about his plans and wrote an opinion piece about it for *The Seattle Times,* warning of what would happen to the lake, its trout, its surroundings, wildlife, and the existing state park if the project went forward.

That story, for which he received no payment, got a huge amount of mail, with scores of letters pouring in from as far away as Pennsylvania. It also triggered establishment of a grassroots "Save Pass Lake" movement headed by an energetic young man who rallied the public to save the lake. For his part, the developer threatened Steve with a libel suit, but it was a hollow threat, because truth is an absolute defense against libel, and the story was true.

The collective efforts of the "Save Pass Lake" movement finally gained the attention of state government, which eventually appropriated the funds necessary to buy the threatened property and protect it by adding it to the state park. Today Pass Lake remains a popular and scenic fly-fishing destination, still capable of producing heavy trout even though the fishing has been degraded recently by the illegal introduction of scrap fish by bait fishermen. Maybe time for another story.

In this case Steve had the satisfaction of knowing he'd written something that made a genuine and important difference, but the more you write, the more you'll discover it's often difficult to tell whether your part in any ongoing conversation or controversy is having a measurable effect. However, if your article is effective—as in Steve's case—pretty quickly a lot of voices and forces will get involved in a movement. You may end up not knowing if your part in the process made a difference, much less *the* difference, but if your article started the conversation you can feel mighty good about it.

Paul had an extreme example of having no idea if he'd done any good early in his writing career. In 1979, he wrote a long article in what was then called *Rod & Reel* magazine (later to become *Fly Rod & Reel*, now unfortunately defunct) about the peculiar legal and philosophical challenges of allowing and managing fishing in national parks. Using four "case studies" from Yellowstone, the Great Smoky Mountains, the Everglades, and the Sierra parks of California, he laid out the little-appreciated history of the evolution of national park policies and how they affected, or could affect, or should effect, angling in those parks' famous waters. Not long after the article was published, John Merwin, then-editor of the magazine, mentioned to Paul that some office in the National Park Service had requested reprint rights, which were of course granted. A few years later Paul used part of the article in a book, but otherwise that was the last he heard of it.

Fast-forward 32 years, to the spring of 2011, to New Orleans and a big conference held by the George Wright Society, a leading professional natural-resources management organization. Paul, who has written more books about natural history and national parks than he has about fly fishing, was there to be honored with the society's communications award, presented for his, "outstanding contributions to conservation history, national park policy, and understanding of wildlife, which have significantly advanced science-based resource management."

Of course Paul was thrilled and daunted by this honor, but what topped it was the award presentation made by the distinguished marine ecologist Gary Davis. Dr. Davis began his comments with a little story, explaining that in the late 1970s he was doing research in Everglades National Park when the park and various interest groups were bogged down in a controversy over the management of Florida Bay. Then along came this article by some guy named Schullery, which the park reprinted and distributed to the various factions involved in their controversy. To Paul's surprise, Dr. Davis then said the article provided the informational impetus needed to break the political logjam so that the controversy could move along to its next stage (such controversies never end). Accolades like this can keep a writer going for a long time, especially if, like Paul, he has no idea if what he's doing matters. If it hadn't been for the award, Paul never would have known about the positive effect of his 32-year-old article.

Humor is another valid reason for writing; there's too little of it in modern fly-fishing literature. If you have a good, funny story—especially if it's true—don't hesitate to tell it. Just make sure you keep things in good taste and don't libel anyone. It's especially easy to avoid libel if the story is about you. Many fly-fishing writers, including no less distinguished a master of the genre than Nick Lyons, have often shared hilarious tales of their own angling mishaps. Such stories are not only sympathetic and entertaining, they also display a humility that is sometimes in short supply in fly fishing.

If you want to write for no other reason than your own satisfaction, there's nothing wrong with that, either. It's hard to beat the gratification you'll feel with acceptance of your first magazine story or book contract, let alone the even greater satisfaction that comes when you first hold the published results in your hands and turn the pages—*your* pages. Few things bring a greater feeling of accomplishment.

Quite a few fly fishers, especially older ones, rightly feel a need to reflect more or less publicly on their experiences, and may not see such writing as a career or care in the least about making money from it. It's just their way of celebrating what they've gotten to do as anglers, and a sort of firming-up of their memories of years astream. Memoirs of this type, which often are the only book the author ever wrote on the subject, are a special, admirable part of the literature. Many writers already prominent in the larger literary world—essayists, novelists, and nonfiction authors whose entire careers have been devoted to other subjects—have eventually felt moved to sit down and express their affection for fly fishing in a book or two. Among those that come to mind right off are Pulitzer Prize–winning historian and biographer Odell Shepard and novelists Craig Nova, Thomas McGuane, and William Humphrey.

Another hopeful reason for writing is that you just might discover a whole new passion, an epiphany that will change your life, while simultaneously enriching the lives of the many people who read your work. It will also add immensely to the satisfaction you receive from fly fishing. You'll never know unless you try.

Why Write: Some Less Good (or Bad) Reasons

There are, however, some motivations for writing that aren't so valid. Money is one, but you need look no further than the advice given in the book we've already mentioned, the first printed English-language book on angling, *A Treatyse of Fishing wyth an Angle*, published in 1496: "You must not use this artful sport merely for the increasing or saving of your money, but mainly for your enjoyment and to procure the health of your body and, more especially, of your soul."

That's clear advice, and whoever wrote those words was right on target. But even if you ignore the venerable and eminently sensible advice offered by the *Treatyse*, you'll find money is usually still a poor motive for writing fly-fishing stories. That's because there's very little money to be made. It does happen, very rarely, that the author of a fly-fishing book rises above poverty-level income; we're all familiar with such examples as *Selective Trout*, *A*

River Runs Through It, and a few others that achieved commercial best-seller status or even reached Hollywood. But it is difficult to overstate the rarity of such exceptions to the rule; they are Cinderella stories, appealing to all writers' hopes and dreams. By all means, feel free to set your sights as high as you like, and more power to you, but don't let coming up short of those heights ruin the joy of the writing or your sense of accomplishment and satisfaction at having contributed to a literature you love.

And don't quit your day job.

The average fly-fishing magazine used to pay $100 to $600 for a freelance story, but since the internet siphoned away most print magazine advertising—and some magazines along with it—some publications have reduced their rates to as little as $50 at the low end or $400 at the top. That's very little money for a lot of work. Online magazines are such a recent phenomenon they have yet to establish a reliably average pay policy.

The average print magazine also probably doesn't buy more than three dozen stories a year out of hundreds or even thousands submitted, so your chances of making a sale are not much better than winning the lottery. Most magazines—both print and online—also are at least partly staff-written, which raises the odds even further.

Of course, you could write a book instead, but the average fly-fishing book published in this country rarely sells more than 2,500 copies, and the author inevitably ends up earning only a tiny fraction of the minimum wage.

It's worth saying again: As you consider your future as a fly-fishing writer, keep this harsh reality in mind: There's very little money to be made in this business. Maybe we should start lobbying for a federal minimum wage for fly-fishing writers.

Another selfish motive for writing about fly fishing is desire for fame or immortality, but those who write for that reason also nearly always end up disappointed. Consider: If you do beat the odds and get a story published in a magazine, the most you can hope for is that some people will remember it, and maybe even remember your name, until the next issue of the magazine is published or goes online, after which both your story and your name will be forgotten. If you think that's wrong, then try to remember the author of a story you read in a fly-fishing magazine a year ago. Unless the author was somebody already famous, we'll bet you can't do it.

Lee Wulff once said the only way writing can make you famous is if you write books, and he was right. But Lee took no chances; in addition to books, he also wrote for magazines, made films, appeared frequently on television, and gave demonstrations at scores of outdoor shows and conventions. He did indeed become famous, but he had to work very long and hard to get there. And very few others have ever achieved such status.

If the practical unlikelihood of achieving fame in the world of fly fishing isn't enough to bring you to your senses about trying to become famous, maybe you should consider that such an ambition is actually against the sport's long tradition of a quiet, contemplative way of life: Really? You want to write so you can become a famous fly fisherman? How embarrassing for you.

Sad to say, though, you won't be alone. In his acclaimed fly-fishing memoir *The Longest Silence*, novelist Thomas McGuane ably summed up your strange ambition by describing certain elements of modern fly-fishing commerce as "an era when famous fishermen scramble to name flies and knots after themselves with a self-aggrandizing ardor unknown since the Borgia popes." If that's how you want to be seen, we're sorry for you.

Another highly suspect motive for writing articles for fly-fishing magazines is to get free trips. Above the constant buzz of sound during commercial fly-fishing shows we've overheard some conversations in which so-called outdoor writers prostituted themselves shamelessly begging free trips from the owners of fishing lodges, promising a highly favorable report even if the trip turned out to be a disaster.

Unfortunately, that kind of thing is all too common, and you don't have to look very far to find the evidence. If a magazine publishes a highly flattering story about some fishing resort, and the author was the magazine's editor or someone on its staff, and the resort just happens to have a full-page color ad in the same issue, you'll know the fix is in. The people who do this might get to fish a lot, and for all you know they're actually giving you the straight story about their experience, but you'll never know for sure, will you?

Why Write: For the Love of it

Most of us write just because we love it, we want to help others enjoy the sport, or we just want to satisfy ourselves. When his first book was published, one author friend said, "this book is my monument." He was right; what we write becomes our monument, the legacy we leave behind after our last cast.

The aim of this book is to help you learn to write for magazines or books, and write for your own best reasons. But before you try writing anything at all, you need to do lots of reading.

Starting with the next chapter.

The first dividend to be derived from reading the old angling authors is the realization that there is nothing new under the sun.

—Arnold Gingrich, *The Well-Tempered Angler*

CHAPTER TWO

Read Before You Write

There is much for fly fishers to be proud of in our sport's literature. We have an almost inexhaustible wealth of intriguing, entertaining, inspiring, and informative reading available to us. But we shouldn't therefore assume the rest of the world feels the same way about us and our literary pretensions, much less imagine that because our sport is wonderful, our writing about fly fishing automatically or necessarily makes us good writers.

Fly fishing's self-perceived literary eminence, such as it is, is so precarious that if you remove only one author, Izaak Walton, from our sport's huge bibliography, we pretty much fall off the scope of most serious scholars of notable English literature. (Walton's fly-fishing writing wasn't much anyway, neither lengthy nor even the result of his own thinking.) Remove a couple dozen others and we lose much more of our claim to literary distinction. We rightly love our books, and we justifiably honor our long literary tradition, but so do gardeners, duck hunters, dog fanciers, baseball fans, photographers, falconers, cooks, birders, and many other passionate specialists about their chosen pursuits. Let's not overdo it; after all, who really cares except us?

And let's also be sure we don't tarnish it. In 1978, in his little fishing memoir *My Moby Dick*, the distinguished novelist, occasional literature professor and avid sportsman William Humphrey described his own reading of the fishing books of that year in a paragraph that should be required reading for all ambitious fly-fishing writers: "The angler had metamorphosed into the ichthyologist, and the prevailing prose reflected the change—if mud can be said to reflect. I found myself correcting it as I had done freshman themes in my years as a professor. You had to hack your way through it as through a thicket. Participles dangled, person and number got separated and lost, clichés were rank, thesauritis and sesquipedalianism ran rampant, and the rare unsplit infinitive seemed out of place, a rose among the nettles. Yet, instead of weeding their gardens, these writers endeavored to grow exotics in them: orchids, passion flowers. Inside each of them was imprisoned a poet, like the prince inside the toad. What came out was a richness of embarrassments: shoddy prose patched with purple—beautifully written without first being well written."

These weren't the words of some grumpy old guy who simply disapproved of everything; Humphrey's book was in fact dedicated to Nick Lyons, certainly one of our best-loved writers (and a former student of Humphrey). Humphrey was also champion of unfairly neglected angling classics, such as Odell Shepard's eloquent paean to trout fishing, *Thy Rod and Thy Creel* (1930).

By the way, Steve and Paul are reasonably sure they know to which writer's works Humphrey was referring in his remarks, but it would be poor sportsmanship on our part to tell, even though by telling you we'd be giving you some books to read to learn how *not* to write.

With that preface, and keeping in mind a reasonable appreciation of our literary tradition moderately tempered by Humphrey's memorable little diatribe, we are ready to consider what it takes to write well, or at least better, about fly fishing.

As we've already pointed out, fly fishing, like writing, is almost always a lonely, highly individual pursuit. But as the great angler-author-conservationist Lee Wulff once put it, fly fishing is also "the most social of the solitary sports." When we leave the stream and are back in the car, and from then until we once again wade into the stream, we find ourselves engaged in an incredibly rich, endlessly absorbing, occasionally quarrelsome, and—let's face it—sometimes really boring conversation. A stranger coming new to the whole enterprise of fly fishing would no doubt notice many things, some good and some bad, about the sprawling community of individuals who make up the sport's enthusiasts, but it's a sure thing that very soon that stranger would have to ask, "Don't these guys *ever* shut up?"

No defense is necessary for our talkative habits. Most of us would have it no other way. Many of us would lose much of our interest in the sport if it weren't for all that stimulating, entertaining swapping of stories, insights, and—again, let's face it—half-baked theories. It's just who we are.

And it's who we've always been. It's safe to assume that long before printing presses, telephones, films, blogs, podcasts and especially all the blessings and curses of the internet empowered us to accelerate wildly the pace of our conversations, anglers must have gathered at their local tavern, club hall, dinner table, riverbank, or shady boulder to brag, listen, hypothesize, lie, celebrate, lament, and otherwise thrash out the day's triumphs and misadventures astream. Whether remembering or anticipating, fishing just demanded a conversation.

Write, Write, Write

Talking has always been good enough for most anglers, but for 500 years now, and for an ever-growing number of us today, that impulse to talk naturally translates into a heartfelt need and ambition to make the words last longer: to write about it all. We can't help it, nor should we try.

The whole point of this book is to encourage and help you do just that: By all means, write it all down. Many anglers keep diaries with no intention ever of doing more than recording experiences for themselves, but many others want to write for a wider audience (we'll have more to say about the ways a diary can help your writing in chapter 6). You must be one of those or you wouldn't be reading this book. Do a good job and you're sure to find readers and become part of the sport's inexhaustible published conversations.

But before you do that, take care. You owe it to yourself, and even more to your potential readers, to know whether you're just repeating what many others have thought and felt and said before. To put it bluntly, if you want to be a fly-fishing writer, it's vitally important to have a good idea of what others have already written. The best way to be a good writer is to be a good reader. Never stop reading. Read the old books to know how we got here, and read the new books to learn what matters most to today's readers.

To his chagrin and eventual amusement, Paul learned this lesson very early in his fly-fishing career. One day 50-some years ago, after a couple years tying his own flies, he "discovered" that by twisting a normal strand of floss tightly into a dense, hard little cord and then wrapping it onto the hook like that, he could create a beautifully segmented body; *Eureka!* Being functionally illiterate about the sport, he had no idea how really old, not to say intuitively obvious, this technique was. Being stupendously naive about how to submit an article, he immediately sent off one of his wondrous new flies and an accompanying letter to the fly-tying editor at one of the magazines, apparently assuming the editor would immediately urge him to write something about this earth-shattering innovation. *An actual publication! Next stop fame!* The editor, bless his heart, thanked him, didn't call him an uneducated hick (which he was), and patiently explained that rolled floss bodies had been around for ages, but thanks anyway. Paul has always been grateful for that kindness, and didn't need to learn that lesson again. Do your homework.

Despite the modern flood of books, magazines, and digital communication, it is still true that your best hope for having a clue about the deeper moods, trends, and traditions of the sport is to explore what anglers have said, thought, cared about, and argued over long before the internet was even a glint in the eyes of a few visionary geeks.

Don't confuse this with some lame homework assignment from your high-school history class; it's a legitimate part of the fun and fulfillment to be found as a fly fisher. With a little luck, you'll find yourself happily exploring this great literary heritage of fly fishing's wonders, an almost endless amount of reading just waiting to be enjoyed for the rest of your life. It can be that good. Don't deny yourself. As the late Alfred Miller, who wrote for many years under the pen name of Sparse Grey Hackle, famously put it, "Some of the best fishing is done not in water but in print." In the rest of this chapter, as a way of starting you on that literary exploration, let us take you fishing in print.

And please: don't imagine we're saying you need to read all this stuff before you start writing. What we're suggesting here is less demanding, but still quite exciting. Exploring fly fishing's grand literary tradition is a surprising, entertaining and endlessly informative part of the sport. In a way, you're luckier than any previous generation of fly-fishing readers, because many if not most of the sport's literary milestones can be downloaded from Google and other websites, usually for free. Those of us who waited many years to get a look at this or that rare book envy you more than you can imagine.

So read what you can when you can. Search out the books recommended by or cited in other books. The more you engage in fly fishing's 500-year published conversation, the better prepared you'll be to take part in it yourself.

Read, Read, Read

We'll start with a book we don't necessarily even recommend to new readers, but it helps set the stage. In 1883, two of the premier British authorities on angling literature, Thomas Westwood and Thomas Satchell, produced what is still one of the sport's most extraordinary and useful books, *Bibliotheca Piscatoria, A Catalogue of Books on Angling, the Fisheries, and Fish-Culture, with Bibliographical Notes and an Appendix of Citations touching on angling and fishing from old English authors.* The extended title hints at the exhaustive nature of their effort, which was in fact a literary treasure thinly disguised as a masterpiece of exacting scholarly form.

Hardly anyone discussing this book today refers to it as *Bibliotheca Piscatoria.* We just say "Westwood and Satchell." They were by no means the first to undertake the job of compiling such a set of lists, but their book was orders of magnitude more thorough than any predecessor's efforts, and remains one of the essential standard reference works among serious angling bibliophiles and general readers. It is perhaps the least read of the sport's genuine literary classics.

Westwood and Satchell serve us well here for its sense of the scale of angling literature, and makes the point that when we say the enjoyment of the sport's literary heritage is a lifelong pursuit, we're not exaggerating. Even that long ago, it took these two scholars more than 400 pages of fine print to fully cite all the fish- and fishing-related works they knew of, virtually everything that appeared after *A Treatyse of Fyshing wyth an Angle* debuted in 1496. In the years since Westwood and Satchell, the pace of publishing on angling subjects has increased severalfold. We really do never stop fishing in print.

We don't mean to intimidate you by telling you about Westwood and Satchell, but to present it as an encouraging invitation to the whole enterprise of becoming a reasonably well-read angler. However, that book does ask the big question: In such a mountain of stuff: Where should you even start? Luckily, there are some easy answers.

For one thing, there are many wonderful "samplers" that give you a good taste of the variety of styles, approaches, and ideals of fishing writing, both modern and historic. These include anthologies of the works of fishing writers generally, and of fly-fishing writers in particular. Some are specialized in one way or another. A few examples serve here. *Astream: American Writers on Fly Fishing* (2012), edited by angler/literature scholar Robert DeMott, is a grand collection of stories by modern novelists, poets, journalists, and other writers, most of whom are not normally associated with fishing writing. *American Trout Fishing* (1966), edited by Arnold Gingrich, presents an all-star cast of the best-known fly-fishing writers going back as far as half a century earlier. Other books focus on a single species of fish, a single style of fishing, or any other presentable topic. Stephen Sautner's *Upriver and Downstream: The Best Fly-Fishing and Angling Adventures from the* New York Times (2010) collects a generous selection from that newspaper's popular "Outdoors" column. Henry Hughes' *The Art of Angling: Poems about Fishing* (2011) presents many of the best fishing poems from the past few centuries, and his *Fishing Stories* (2013) ranges more broadly than most such anthologies into international waters for old tales from other literatures than English. Holly Morris' *Uncommon Waters: Women Write about Fishing* (1991) and *A Different Angle: Fly Fishing Stories by Women* (1995) suggest the extent to which women have been and now are an important part of fly-fishing culture.

There are many more anthologies, old and new, but among all these choices, a longtime favorite of ours, and still one of the best, was compiled by our friend Nick Lyons about 50 years ago: *Fisherman's Bounty: A Treasury of Fascinating Lore and the Finest Stories from the World of Angling* (1971). You can't go wrong with this book, as it ranges through the long publication history of the sport, selectively hoovering up the best stuff throughout. It seems to be out of print now, but thanks to the internet, "out of print" no longer means "unavailable," and with very little effort you can find reasonably cheap copies of it—and, by the way, most of the other books we'll mention here. Nick's anthology is worth the price of admission just to get to read Elizabeth Bishop's poetic masterpiece "The Fish" and Rudyard Kipling's hilarious tale "On Dry-cow Fishing as a Fine Art." You might also try Nick's more recent (and even larger and more diverse) anthology, *The Gigantic Book of Fishing Stories* (2007).

There is another such book, though, which stands out not only for its rich assortment of writing from many important historic and modern authors, but also for its helpful and entertaining commentary. It's Arnold Gingrich's *The Fishing in Print: A Guided Tour through Five Centuries of Angling Literature*. Originally published in 1974, used copies are still easily and cheaply available online.

(Important note to publishers: Nick has since produced an even grander anthology, but Arnold's book should be reprinted immediately, in cloth, paper, and e-book editions.)

Arnold, who borrowed the title of his book from the earlier-quoted statement of Sparse Grey Hackle, was himself something of a phenomenon in angling publishing. Founding editor of *Esquire*, through whose long, colorful history he published famous writings by many of America's leading literary figures of the mid-1900s, Arnold was a happily passionate amateur among the fly-fishing experts. Extremely well-read, he was well off enough to be able to afford many of the rare old books long before they became available to all of us through the internet, and was furthermore a consistently companionable (a word he used to describe his favorite fishing books) writer. He was thus perfectly positioned to produce this long, hospitable book in which he quoted many historic angling books at great length while providing abundant commentary on the worth and context of each. Beyond feeling obligated to tell you that the accuracy of his commentary has become a bit dated in a very few specifics, we wholeheartedly recommend not only *The Fishing in Print* but also his earlier memoir, *The Well-Tempered Angler* (1965), which, amidst tales of his angling adventures with many of those same literary greats he just then was publishing in *Esquire* (including a chapter titled "Horsing Them in with Hemingway"), constantly revisited the sport's early and recent literature. He even devoted a chapter to some excellent lists of historic titles that he regarded as essential reading. We don't disagree with any of his choices.

The other way to get started is to read fly fishing's history. There have been many British and American books on the subject, from general surveys such as Paul's *American Fly Fishing: A History* (1987) and Andrew Herd's magisterial *The History of Fly Fishing* (get the third edition, 2019), to a host of more narrowly focused books. Herd's book ranks by far as the most thorough history of the sport ever undertaken, with the added bonus of being much more entertaining than any high school or college history class you might remember painfully. Now that it's a quarter of a century old, Paul's *American Fly Fishing*, though still the prevailing overview of the subject, is showing signs of its age in its treatment of some historical details we now understand better than we did, and of course it lacks any coverage of the sport's most recent, energized quarter century. But it has been regarded highly enough to go through at least five editions, so it must still be pretty good. (Paul is safe from accusations of greedy self-advertising here because he gets no royalties from any of these past editions, new or used; from the date of its first publication, all royalties have gone to the American Museum of Fly Fishing.)

These books are only the beginning for any enthusiastic reader of fly-fishing history. We'll discuss some of the possibilities for writing on historical subjects later in this book, but here we should at least point out that the history, natural history, and angling lore of many—and more all the time—of the best-known trout waters have been researched and written up in book-length studies. To name only a few of the most famous American rivers that are the subject of books, the Beaverkill, the Neversink, the Battenkill, Michigan's Pere Marquette and Au Sable, Arkansas' White River, Maine's Grand Lake Stream, the Madison, the Henrys Fork, the Yellowstone, the McKenzie, the North Umpqua and many others are now thoroughly chronicled in books aimed at anglers. Even more streams in the UK, where anglers have been fishing and studying their favorite waters for a much longer time, have been the subject of books.

The current spirit and ideals of the sport of fly fishing are largely the result of its long, celebrated history. Generations of fly fishers have found their experiences of fishing a given stream enriched by knowing who fished there before them, how they fished, and what they thought of the place. For just a couple examples, in the East anyone fishing Pennsylvania's Letort will enjoy it all the more for knowing the stories and reading the writings of George Gibson, Vincent Marinaro, Charles Fox, Ed Koch, Ed Shenk, and all the others who preceded them there. In the West, it's impossible to appreciate the tradition of fishing Montana's Madison River without the writings of Howard Back, Charles Brooks, Craig Mathews, Bud Lilly, Ted Leeson, and their compatriots.

KNOWLEDGE AND POWER

Henry Van Dyke, Princeton professor, U.S. ambassador to the Netherlands, distinguished clergyman, and an enthusiastic fly fisher (*Little Rivers*, 1895, and *Fisherman's Luck*, 1899), divided outdoor writing into two types: the literature of knowledge and the literature of power. The first includes the instructional manuals that form the overwhelming majority of fishing books. The second includes the meditative and inspirational, most often presented in the form of stories about the personal experiences of the author.

Now we routinely refer to the first type as "how-to," and we often refer to the second type, with all its explorations of the aesthetics, moods, personalities and philosophies of the sport, as "why-to." But for at least a couple centuries there has been one other very common type, kind of a subset of the how-to that has become its own huge literature, the "where-to." Modern fly-fishing magazines, books, and web pages are full of spectacularly photographed stories of many exotic destinations for us to dream about.

Like most sweeping generalizations, these categories aren't perfect, nor do they really encompass all the types of fishing books. Among other things, they leave out the natural-history books, art books, and conservation-oriented books, all of which fit in our definition of books of intense interest to fly fishers (exactly where fiction fits into this picture depends on the fiction, which may include elements of all of the above; same goes for poetry). The truth is, there are very few fishing books that are exclusively devoted to only one how-why-where topic; we just don't write, think, or fish that way. The driest fishing manual may surprise us with a few pages of personal anecdotes to illustrate some important point, or with an unexpected sermonette on conservation or good fishing manners.

1496 to 1800: No Fly Fishing Only

For the practical purposes of a new reader of fly fishing books—call it angling literature if you like the higher tone of that—keep in mind that for the first 300 years of fishing-book publishing, there were no English-language books devoted entirely to fly fishing. This is one good reason why many of the best anthologies do not restrict themselves to fly fishing; it has always been true that many of the best fishing stories are about bait- and lure fishing. From the appearance of the *Treatyse* in 1496 until the first years of the 19th century, fishing books were almost always devoted to whatever kinds of fishing were most familiar to the author; sometimes fly fishing was dealt with in considerable detail, sometimes only in passing. For two centuries after its first publication, the *Treatyse's* brief list of 12 fly patterns (famously known as the "jury of 12") was regularly plagiarized almost verbatim by later writers, especially those who didn't know very much about fly fishing to begin with. This might suggest that fly fishing was an essential part of being a well-rounded angler; most of the experts who wrote fishing books may have at the same time regarded it as a comparatively minor part of the sport, not one they had personally experienced.

Even so, it seems likely the biggest reason for a lack of books devoted exclusively to fly fishing is that fishing, like any other human pursuit, is always evolving, and most anglers had not yet begun to see fly fishing as anything other than one of several techniques available to them if they wanted to be successful. No doubt there have always been people who preferred mostly or exclusively fishing with flies—some of the earliest European tracts on the sport suggest just that—but most anglers back then were generalists who could ably use the same rod to fish with bait, lure or fly as the circumstance dictated.

The full story of how fly-fishing writing came into its own is no doubt more complicated than any single historic source could show, but in the early 1700s we do begin to see clear published evidence not only that fly fishers saw their version of angling as distinct from the others, but also that some fly fishers even saw themselves as special, or even superior to their fellow anglers. The poet and playwright John Gay, in his *Rural Sports* (1713), versified at some length about why fly fishing was a higher and even more moral form of fishing because it did not involve killing various small animals for bait: "Around the steel no tortur'd worm shall twine/No blood of living insect stain my line." In his self-congratulatory elevation of fly fishing, Gay conveniently forgot about all the creatures whose often cruel deaths were required to supply him with the wealth of materials used to tie his flies, just as he neglected to consider the tortur'd *fish* whose blood was spilled when he caught them. But then who said that fishermen have ever been careful thinkers, especially when they're on a rhetorical roll?

After Gay's time fly fishers became more vocal about the imagined superiority of their sport, but that was hardly the most important reason that by the early 1800s fly fishing began to generate books of its own. A whole array of factors—improving printing and engraving technology, advances in tackle construction, the increasing number of recreational anglers, a world trade that brought a vivid assortment of exotic fly-tying materials from the British Empire's possessions in the far corners of the planet—led to a dramatic increase in the rate of publication of all types of fishing books throughout the 19th century. Perhaps more important, fly fishing writing, like all other writing, was caught up in the prop wash of that greater increase in publishing of all kinds during the Industrial Revolution.

Fly Fishing Writing Comes into Its Own

This is no place for an exhaustive review of even the most notable of the first fly-fishing books, but there's no harm in naming a few milestones from such a dynamic period. Starting with George Scotcher's *The Fly Fisher's Legacy* (about 1810) and George Bainbridge's *The Fly-fisher's Guide* (1816), fly fishers went on to develop a formidable, informative, and (no surprise here) occasionally quarrelsome library of their own quite separate from the generalist-anglers. Though Scotcher, Bainbridge, and others before them had discussed trout-stream insects to one extent or another, Alfred Ronalds' *The Fly-Fisher's Entomology* (1836) established a durable gold standard of careful scientific thinking; many of Ronalds' insights can still be read to advantage, including his modern discussion of how light's passage through water affects trout and anglers' perceptions of one another. Ronalds' book offers us clear evidence that almost 200 years ago there were savvy anglers concerning themselves with such seemingly esoteric but vitally important matters as the physics of light.

Though fly fishers had been attempting to imitate insect life—and other stream creatures—at least since they started writing about it, the development of modern angling entomology is a good case study in how our literature both advances and reflects the sport's restless fascination with every aspect of fly-fishing's subtler details. Post-Ronalds studies of angling entomology through the rest of the 19th and the early 20th centuries resulted in progressively more advanced British books—by Francis, Halford, Skues, Mosely, Mottram, and a host of others—though it was upward of a century (99 years, to be precise) before American anglers got past dabbling in the subject and successfully produced a regional fly-fishing entomology of high scientific precision, Preston Jennings' *A Book of Trout Flies* (1935). Jennings' book heralded the arrival of ever more refined studies of regional or national application—not only of the insects but of our enthusiastic efforts to imitate them better. Again there's a litany of names—Flick, Wetzel, Grove, Marinaro, Schwiebert, and others up to and including Doug Swisher and Carl Richards' epochal, best-selling *Selective Trout* (1971), and on from there. (By the way, the first American book devoted exclusively to fly fishing was George Dawson's *The Pleasures of Angling with Rod and Reel for Trout and Salmon*, published in 1876, but after that there was a growing stream of fly-fishing titles).

Don't shy away from those old generalists, though. Today's literature-minded angler should still want to read the early generalists, not just because of their fly-fishing material but because of everything else they say about the fishing of their times. Among American fly fishers, no 19th-century writer was held in higher regard than Thaddeus Norris, whose colossal *The American Angler's*

Book (1864), not only contained excellent guidance for fly fishers, but also covered every other part of American angling so expertly and sympathetically that Arnold Gingrich called Norris "the American Walton." Among Americans, Norris' book was to the 19th century what Ray Bergman's *Trout* (1938) would be to several generations of 20th-century anglers.

THE BOOK OF THEM ALL

It must be said that everyone should give Izaak Walton a try. It is, after all, the only fishing book to have long ago achieved near household-name status among non-anglers, and it can start you in many reading directions.

Some purists say the very first edition of Walton's *The Compleat Angler or the Contemplative Man's Recreation* (1653) is the one to read, that Walton got his story just right first time around and became a little verbose in the subsequent editions. But the majority of us agree instead that you'll get the most out of the last edition published during Walton's lifetime, that being the fifth (1676). The great advantage of the 1676 edition for fly fishers is that it's the first to include Charles Cotton's milestone essay, "Being instructions how to angle for a trout or grayling in a clear stream." Cotton's contribution was mostly devoted to fly fishing, about which he was vastly more informed and practiced than Walton, who had more or less just cadged information from the *Treatyse* for his fly fishing-discussion in previous editions of the *Angler*. Because it combines the work of Walton and Cotton, the 1676 edition is by far the most often reprinted.

A. J. McClane, in his immense *McClane's Standard Fishing Encyclopedia* (1965), summed up both the challenge and joys of reading Walton today: "New readers to *The Compleat Angler* are usually bewildered by the fact that his book is written in dialogue form. We expect a direct monologue in our angling works, but in Walton's day there was time to wet now-dried Elizabethan prose, and it is a pity really that words are perishable. As the two men walked to their river, the tremendous secret of an honest life was unfolded. Izaak's literary harvest was merely the shreds of these conversations, for he was by nature a profoundly inquisitive man. Even in reading him as the practical fisherman we find none of the dull stuff chronicled by scholars in the years following. To this day, nobody has written a more finely detailed study in the art of using live baits. Walton's eloquence lagged only when the subject of fly fishing became too pressing, and here he turned to his friend, Charles Cotton."

Such a statement from McClane, surely one of the 20th century's foremost expert how-to anglers (his own book on fly fishing was, after all, called *The Practical Fly Fisherman*, 1953), should be some indication that, for both inspirational and instructional reasons, Walton might be hard work but is important reading. In *The Habit of Rivers* (1994) Ted Leeson, one of today's smartest and most literate fly-fishing writers, has said he "would rather chew little balls of tin foil than reread" Walton. You might come to that conclusion too, and maybe it won't prove you're an insensitive brute, but notice that Ted said "reread," so apparently he did get through Walton's book at least once.

A curious and appealing distinction of the 1676 edition of the *Angler* is that you could buy Walton's Part 1 as a separate book. Or you could buy it in combination with Cotton's Part 2. Or you could buy Cotton's Part 2 as a separate book. Or you could buy the two in combination with another important 17th-century book, Colonel Robert Venables' *The Experienced Angler: or angling improved*, which had first been published in 1662 and was in its fourth edition in time to be combined with Walton's and Cotton's books in a grand volume titled *The Universal Angler*. The publisher was keeping shoppers' options open by providing all these alternatives.

There have now been more than 400 editions of *The Compleat Angler*, with more appearing all the time. That's not merely because it was a milestone of fishing literature, but perhaps more important, because it is "a transcript of old English country life, a study of the folk heart," as the novelist-diplomat-angler John Buchan put it in his introduction to the Oxford University Press edition in 1935. In the view of the past couple of centuries of English literature scholarship, Walton's gifts as an angler were nowhere near as significant as his gift for evoking a pastoral idyll.

Though none achieved fame beyond angling circles, there have been many such all-round fishing books since Walton's *Angler*. A representative very few of the best would include Richard Bowlker's *The Art of Angling Improved in All Its Parts, Especially Fly Fishing* (1746, with several enlargements in later editions by Richard's son, Charles), and Francis Francis' *A Book on Angling* (1867 and later editions) in the UK, and Thaddeus Norris' already-mentioned *The American Angler's Book* (1864) and Ray Bergman's *Trout* (1938) in the United States.

Translating from English to English

Reading the early fishing books might remind you of the old saw that Americans and British are two peoples "divided by a common language." The further back you read in fishing literature the more that will seem the case historically, too. Even books published 100 years ago can have some pretty formidable and obsolete jargon, and the 300-year-old ones can at key points seem impenetrable.

But they're not. In most cases you can figure out what they're talking about just from the context, and if not, Google awaits with definitions of many obscure terms. It's part of the fun of reading the older books to mentally hear their conversations in their own voices, as jarring as they might sound at first. As McClane (quoted above) suggested, the different pace of the texts, the puzzling arcane terminology, and the unfamiliar British settings of the older books can require some adjustments.

Some of the most important books (including both Walton's and Cotton's) that were written in dialogue, and many more that weren't, may seem tediously stilted to the modern ear. Walton began chapter 1 of his fifth edition with one of his characters, Piscator by name, saying, "You are well overtaken, gentlemen, a good morning to you both; I have stretched my legs up Tottenham-hill to overtake you, hoping your business may occasion you towards Ware, whither I am going this fine, fresh May morning." There's lots of this sort of talk before they get around to any serious fishing, but remember that for as much as any other reasons, you're reading for immersion in another fishing time and place; trust us, there will be plenty of other early fishing books whose texts are so aridly businesslike that you'll begin to wish for a few whithers and fresh May mornings (when it comes to describing fly patterns, almost none of these early authors gave us sufficient detail to tie a fly with confidence that it looked much like the ones they were using, thus leading several generations of more recent fly tyers into irresolvable debates about just the right way to tie this or that pattern).

Our best advice when reading prose that seems at first glance so strange is just to soldier on; you'll pick up the rhythms and meanings soon enough. A great many very smart readers before you have done it. If you just keep in mind that trout are trout, streams are streams, mayflies are mayflies, and that all these guys were up against exactly the same challenges and mysteries we face today, you'll get along fine and be glad you bothered.

The Modern Writer of Them All

There is no way in good conscience for us not to insist that you make a special effort to read the books of Roderick L. Haig-Brown. For the half-century before his death in 1976, Haig-Brown was foremost among those very few writers whose prose might justly be termed "literature" in that word's richest meaning. Novelist and sporting writer Thomas McGuane, one of the notably literate readers qualified to appreciate Haig-Brown's preeminence in American angling letters, said, "for many who regard angling as the symptom of a way of living rather than a series of mechanical procedures, the writings of Roderick Haig-Brown serve as scripture."

Born and formally educated in England, Haig-Brown first came to western North America in the 1920s, and after some back-and-forth America-England years as he launched his writing career, settled at Campbell River on Vancouver Island, where he and his wife Ann raised their family. He became one of western Canada's most distinguished citizens, a conservationist of national scope, a frequent member of a variety of government commissions, a magistrate, a Canadian army officer, and eventually chancellor of the University of Victoria. But the work that drove him and most defined his life, and that we now remember him best for, was writing some of the 20th century's most eloquent, thoughtful, and enduring books on fishing, conservation, and a life well-lived in close company with the natural world.

For a generous sampler of the writings of Roderick Haig-Brown, you might start with *To Know a River: A Haig-Brown Reader* (1996), edited by his daughter, Valerie Haig-Brown. It's a great collection, and like many of the titles listed here is now in print from the Lyons Press. Haig-Brown's famous four-volume season cycle of fishing books, originally published in the 1950s and 1960s, has been republished as a single book, *The Seasons of a Fisherman: A Fly Fisher's Classic Evocations of Spring, Summer, Fall, and Winter Fishing* (2000). Other especially notable books include *Return to the River* (1941), the dramatic biography of the life of a Columbia River king salmon, and *Measure of the Year* (1950), a memoir of Haig-Brown's life as a farmer, hunter, magistrate, and father living in rural British Columbia. Having recommended all those, we admit that our favorite may be *A River Never Sleeps* (1946), his chronicle of a year of fishing his home river and beyond. But then we've never met a Haig-Brown book we didn't like.

Where to Find Them

Time was when the older, rarer books, especially those published before the early 19th century, were entirely the province of relatively few people: those like Arnold Gingrich with the means to buy them, and those (again like Arnold) who lived near one of the great angling libraries at Harvard, Yale, Princeton, the University of California, and a few others. There were also huge private collections, but they were, indeed, private. Even out-of-print books published in the first half of the 20th century required a certain amount of chasing down. It's only in the past half-century or so that the market in reprints of the older books was recognized by such enlightened editors as Nick Lyons, whose Sportsman's Classics/Crown Publishers editions of numerous important out-of-print titles led the way in getting many great but neglected books back in the hands of new generations of fly fishers.

Those of us who wanted to own some of those older books (after all, the books in big university library collections couldn't be checked out—you had to read them there) did have some occasions for hope. There were countless used-book stores around the country where, with enough patient browsing of dusty shelves (the dustier the better, it sometimes seemed), it was occasionally possible to discover copies of long-wanted books, and there were the occasional catalogs of several used sporting-book dealers. It is difficult to convey, in this age of instant web access to almost everything, how much of a treasure hunt building a modest fishing library was even 30 years ago. For those of us on the bottom rung of the financial ladder, such libraries were often made up in good part of what were known as "reading copies," which was a polite way of saying "all the pages are here, but otherwise the book is battered, stained, marked up, and dismantled to the extent that the only real binding is a rubber band or two." But we had the words, and that was what mattered to us; we weren't book collectors, we were word collectors.

Today if you want to get a look at some rare old book you have options almost beyond counting. There are excellent, authoritative dealers in out-of-print books, in both the United States and the UK, and there are several publishers bringing out new editions of classics, often with helpful modern forewords or other commentary. Or you can resort to the web: Amazon.com, Abebooks .com, and many other sellers can connect you with an unimaginable number of sources. Any web-savvy shopper can easily track down titles we used to wait years to find. In all the above cases, it's easiest and perhaps best just to Google the title and author of the book and see what shows up.

The other way to approach the search is also brought to us by the digital revolution, particularly modern scanning technology. Many and varied business and educational enterprises, more or less beginning with Project Gutenberg about 50 years ago, developed technology to scan books and other documents (Project Gutenberg reportedly began with the Declaration of Independence), a technology that has advanced enormously and, with the enthusiastic cooperation of many of the world's foremost university and public libraries, has made available online many *millions* of long-out-of-print books on every topic under the sun that are often free for the downloading (it's also ignited an apparently permanent quagmire of legal challenges, as all sorts of previously unimagined copyright issues arise in such a free-for-all process).

If you're an avid reader and book lover you might enjoy learning how all this came about. It's quite a saga, featuring the development of devices that can scan an entire printed book in a matter of minutes, hugely improving "optical character recognition" (OCR) technology that literally reads the page as it scans it, turning it into what we mere mortals would call a word-processing document that can be searched, rewritten, or otherwise manipulated. In other words, some ancient text on paper is fed into one end of the machine and a modern searchable e-book comes out the other (almost all the time the original book also comes out, undamaged).

E-Books and Others

For your purposes as a potential reader, all these suddenly available rarities come in two forms. By far the most common is the aforementioned OCR-generated e-book which, depending upon the competence and conscience of whomever created it, may be exactly what you need or it may be a frustrating chimera. In the course of reading and writing projects involving hundreds of these "new" electronic editions, Paul has noticed several common problems with the digital versions of old fishing books.

For one thing, OCR is still an imperfect craft, and it's not hard to sympathize with the machinery however annoyed we get with its operators. The scanner is often dealing with letters printed in type hundreds of years ago: irregularities on individual letters; broken letters; uneven paper surfaces that also may be stained or otherwise damaged; smudges, smears (clumsy fishermen have been handling the book for a long time), fly specks, and other random markings that the bewildered scanner might have misread as commas or other characters, and other similar problems.

Then, whatever proofreading process the scanned files undergo—whether electronic or involving an actual person—just isn't up to the job. Though the text may be perfectly readable for general purposes, you can never count on it being precisely what the author originally wrote.

Far worse things go on in the mass production of these enormous digital libraries. Some of the outfits preparing the books cut corners left and right. The laziest of them may not give you the whole book, or even tell you whether you're getting an abridged edition or the whole thing. As bad, all too many of them simply dump the illustrations, which are so often an essential part of a fishing book; scanning and placing an early book's engravings takes a little extra time, which is evidently just too much bother for the cheap e-book makers. They also routinely delete all the "front matter"—that back-of-the-title page fine print that includes copyright and publication date information, as well as various other credits and details. That information can be vitally important to the reading fly fisher, who needs to know if she/he is reading an OCR scan of the edition that appeared in 1750 or some hacked-up reprint edition published in 1830 with all sorts of changes introduced by various people long after the author was dead.

It is ardently to be hoped that with time these shortcomings of OCR and weak-willed publishers will be cleared up. Until then, you can still enjoy the incredible bounty of access to giant e-shelves' worth of fishing books you otherwise never would have seen. Just keep your eyes open and don't entirely trust anyone. Your best bets by far will be e-books produced by the reputable commercial presses and university presses, which often take great trouble to clean up the little messes left by the scanners, and are not likely to cheat you out of a few chapters or other important things; if they do leave something out they'll usually tell you.

Your other choice is to find a digital version of the book that has been scanned but left in its original layout; that is to say, what you get is essentially photographs of the covers and every page of the book. Thanks mostly to friends who are much more highly skilled at online work than he is, Paul has accumulated several shelf-length equivalents of pre-1920 fishing books like this in PDF format. The best of them even retain the gifts of OCR, so that, for example, if he is reading along and comes upon a particularly juicy passage that he'd like to save, he can simply select and copy it, then paste it into another document for whatever purpose he has in mind.

The big advantage of reading the old books in this format is that you see exactly what the author and original publisher made of the book. All the fine print is there, the illustrations will even have whatever hand-coloring they were given, and as you read you can soak up the period typography and design, which was often quite elegant and is usually more fun and convincing than whatever modern typeface options your computer, Kindle, or other e-reader might offer. Best of all is knowing you're not getting a watered-down, chopped-up, or otherwise altered version of the book. You're reading the book as its original purchasers did.

Happy reading.

I started writing simply because I loved fishing and wanted to share what I learned from my endless experiments with fishing tackle and tactics.

—RAY BERGMAN, *FISHING WITH RAY BERGMAN*

CHAPTER THREE

The Write Stuff

IF YOU'VE ALREADY TRIED YOUR HAND AT WRITING, good for you; the more experience you have, the better. If you haven't started, then it's high time you did.

So—first things first: You'll need a place to write, an office or study or some other space where you can have privacy and quiet. It should have good lighting, either natural or artificial. Natural light is usually better, but as very few of us live in an ideal world where we can choose our time to work, and as fly-fishing writing isn't likely to make enough money to free you from your day job, you may be squeezing writing in when you can, including late at night.

A good, stable writing desk or table is essential, along with a comfortable chair with adjustable height so you can establish the correct angle and distance from your eyes to the computer screen and reach the keyboard easily without having to hunch over or stretch. You're going to be spending lots of time in that chair, so be sure you get it right.

Over the course of our careers we have both accumulated lots of reference books that still serve us well, but the internet also can be an inexhaustible source of reference material if you know how to use it, so you'll need a good, reliable internet connection to search quickly for facts, word spellings, definitions or just about any other information you need.

Since we're talking about the internet, this is as good a place as any to offer a word of caution. Wikipedia and many other online information sources sometimes get things right but not always. People often cite Wikipedia as a source as if they're citing Holy Scripture; don't do that. In fact, using the web's immense quantity of sources is nearly an art form of its own, and is one of the trickiest challenges facing modern writers seeking up-to-date, accurate information. Remember, your writing reputation depends on your accuracy. When you see something in print, especially on a well-designed website (or in a book, for that matter), it provides a powerful, subconscious visual stimulus that you should trust it. But you must look past the cosmetics and do your best to determine if the pretty words actually make sense.

For information of the highest reliability in every field of study backed by a scholarly discipline, you should make the effort to become acquainted with Google Scholar, which weeds out thousands of casually written websites of uncertain reliability and concentrates on an amazing array of academic sources, nearly all of which go through formal peer review prior to publication.

This is also a good place for a little digression on science. Science isn't perfect, but as Carl Sagan once said, the essence of science is that it is self-correcting. Because of the rigorous, self-enforced discipline of science, in both the research and publication processes, the results will be an order of magnitude more trustworthy than popularly written publications. We fly-fishing readers are lucky that we do have many trustworthy popular writers. With luck and careful attention, over time you as a reader will likely come to know which ones they are. You will also discover that the best ones are also the ones who know how to make the best use of the scholarly studies relating to their topics. So please, take the time to learn how to read technical papers and journal articles.

In that vein, here's another tip: In scholarly circles, books, even those published by distinguished university presses, are generally regarded as a little less trustworthy than journal articles. Paul has heard otherwise smart readers say that a book "has so many footnotes; it must be true," when in fact the book in question was just malarkey with the formal trappings of apparently accurate documentation. Still, almost all the time scholarly books on a given subject will serve you better than popularly written books. Just remember the real mother lode of the most consistently reliable and detailed sources of information will be found cited in the book's endnotes and bibliography, the ones that appeared in peer-reviewed journals and monographs.

Speaking of other sources, never underestimate the value of a dictionary, and keep in mind that a computer spelling checker isn't quite the same thing. Just about every "word-processing" program includes a spelling checker, and these are a great help, but they have to be used thoughtfully. For example, Steve has received many annoying unsolicited letters from a real-estate agent who wants to buy his house; the realtor keeps bragging about his career, except he spells it "carrier." His spelling checker didn't flag that as an error because, even though it's spelled correctly, it's not the right word, and spelling and grammar checkers aren't normally sophisticated enough to recognize such errors. A realtor who isn't sharp enough to catch such an error probably isn't going to get much business. Neither is a fly-fishing writer. As the saying goes, "Spell-check is my worst enema."

It's a good idea to keep a modern dictionary and thesaurus close at hand, but most computer writing programs also include a thesaurus, if you can figure out how to use it. Being old guys and admittedly set in our ways, we will always have actual dictionaries and thesauruses handy, but these days for simple needs you can usually bypass the formal dictionary and thesaurus applications simply by typing what you're after, such as, "catadromous definition," or "lateral bifurcation synonym" into your search engine. The same goes for spelling; you can just as easily type in your best phonetic guess at the spelling of an unfamiliar word with good odds that your computer will figure out what you're after and show you the correct spelling.

With some hesitation we recommend using a thesaurus. It's a great tool, but it's easy to fall for the enchantment of all those strange and rarely used words. Beware "thesauritis." The need to write clear, straightforward prose should never be sacrificed in the interest of trying to impress readers with your arcane vocabulary.

A copy of *Bartlett's Familiar Quotations* also might come in handy (and is always fun to browse) but, just as with the thesaurus, be careful about loading up your writing with stuff that you obviously just cadged from *Bartlett's*. In his brief and checkered career grading the papers of college undergraduates, Paul noticed some would write a few pages and then seemed to think, "It must be about time for some pithy quotation," so they'd lift something wholesale from a doubtful website and drop it into the text—clunk—whether it fit or not. And that's exactly how readers read such things—clunk. Quotes that invoke some famous person or a famous aphorism must be used discreetly, infrequently and with a discriminating eye for their relevance, or they lose their effect.

You'll also need some shelf space for those books and others you'll undoubtedly want within easy reach. It might not be necessary at first, but eventually you'll also probably need some file storage for your manuscripts and research materials (and, let's hope, publishing agreements). It's a good idea to set up such files from the start; otherwise, a few years later you'll almost certainly have to waste time going back and trying to reconstruct research notes, correspondence, and so on.

This is the moment when you must admit that writing may be a wonderful soul-elevating art, but it's also a business and needs to be treated as such. Now that almost all such files, including almost all correspondence with magazines and publishers, are generated on your computer, just do the same thing with your computer's filing system. With a lot less effort than we used to expend labeling and maintaining actual cardboard folders, you can create a master folder, say in Word, for each project, and nest as many subfolders in it as are necessary to organize early drafts, research notes, correspondence, contracts, illustration information, business expense documentation, random ideas for topics to include in the piece and whatever else you might need to save.

The same sort of computer filing system can be a very handy way of saving good writing that you discover you don't need for whatever you're working on right then. It happens to many of us that we'll find that while writing on a certain topic we've drifted off into a parallel matter and must delete several paragraphs or pages. It helps ease the pain of feeling like your work has gone to waste if you maintain a folder or two to dump such things, in hopes that they'll be useful later; they often are.

Some writers also like to listen to music while they work (we both do, though Paul is almost completely unable to work with music involving singing because he involuntarily stops working to sing along, which no one on earth should have to hear). If you're among those who like to listen while writing, and if you share your living space with someone who might not share your musical tastes or your conviction that you're a great singer, get a good pair of earphones, which also have the great advantage of reducing distracting outside noises.

Writing Machines

Of course you'll also need something you can use to write. Steve fondly recalls writing his first two books seated in an easy chair cradling a Smith Corona portable typewriter in his lap, but those days are gone forever; we all still have laps, but the typewriters aren't in them anymore. Now everything is done digitally—writing, editing, communicating, contracts, typesetting, royalty statements—you name it. As a writer, it's absolutely essential for you to have a computer and a good word-processing program—if you can find one.

It's very unfortunate that such programs ever got the name "word processing." "Processing" sounds more like something you'd do to make cereal or analyze urine samples. Words are much too important to be "processed;" they are the fruits of our thoughts, created with purpose and care and matched with other words created with equal care and precision, all of which is far beyond the capability of any computer software. Think of it this way: Once in print, your words are more or less chiseled in stone for all future time, so you want to be very careful what you say and how you say it.

Steve well remembers when he bought his first "personal computer" along with a "word-processing" program. After absorbing the features of the program, which took a little time, he began using it to write another book—and that's when he discovered that writing on a computer can be a double-edged sword.

When the book was half complete, he paused to read over what he'd written and discovered major problems; the manuscript was appalling and it was painfully obvious something had gone fearfully wrong. Analyzing the problem soon produced an answer: Back in the days when he sat in an easy chair and wrote on a typewriter, Steve had gotten in the habit of pausing after each paragraph to consider carefully what to say in the next one, because if he made a mistake he'd have to retype the whole page, sometimes more than once. The computer program's ease of doing things had made him lose that discipline, and the results were disastrous.

But what to do about it? Steve ended up rewriting the first half of the book by hand on yellow legal tablets, forcing himself to organize and prepare each paragraph in his mind before committing it to paper, as he had formerly done on the typewriter. It was a tedious process, but it worked; the rewritten manuscript was a great improvement, and when he was confident he had recovered his writing "voice" Steve transferred the rewritten text back onto the computer, then relied upon his newly restored discipline to finish writing the book.

That was long ago. Since then Steve has written 15 more books on computers, always using the same disciplined approach. Now he's on his fourth different "word-processing" program, having been forced to change because of continuing "advances" in software and hardware. But it seems that as these programs become more "advanced," they also become more difficult to use. Not only that, but the so-called advances are poorly documented in any sort of manual, if at all, which often makes it necessary to learn critical functions through frustrating and time-consuming experimentation, or through the risky expedient of getting online and listening in on the countless conversations in which other computer sufferers compare notes on their similar problems. Things seemed a lot simpler back in the days of typewriters. But computers and "word-processing" programs have become the state of the art and it appears they will remain so as we continue feeling our way uncertainly into an increasingly digital future.

Paul's experiences were much like Steve's, except Paul only grudgingly embraced computer culture and maintains that he still has a "hate-hate" relationship with all digital devices. He also wrote his first two or three books by hand with whatever pen or pencil was handy, and on a great variety of types of paper: legal tablets, the blank back sides of discarded photocopies, or other pieces of paper that were only roughly of the same size. He didn't do it this way out of some arcane devotion to what writers had always done; he just didn't know any better, and naively assumed this was what everyone did, not realizing that generations of journalists, academics, novelists, and others had been banging out first drafts on their typewriters for much of the 20th century. Once the handwritten draft was complete, Paul would type it up, using this stage to do whatever additional editing or other cleanup seemed appropriate.

It was a great way to start, though. The discipline of writing entire books with a pen or pencil, one word at a time, tends to produce a concise economy of expression. When Paul finally switched to just writing a first draft on a typewriter the discipline enforced by the handwritten approach apparently stuck, and for good or ill the stuff came out about the same as it had when he was writing it by hand. In fact, the first and only draft of Paul's first big fly-fishing book, *American Fly Fishing: A History*, published in 1987, was written on an IBM Wheelwriter (that's a fancy late-stage typewriter, for those born since 2000) and came out so clean it went directly to the publisher for routine copy editing. But he wasn't always so lucky, or that well organized in his thinking, that he could pull off a one-shot manuscript. Like many writers, the older he gets the more time he spends revising what he's already written.

Which brings us to the world today, and the overwhelming and sometimes troublesome domination of it by the computer industry's 800-pound gorilla, Microsoft.

The program both Steve and Paul use now is the latest version of Microsoft Word. That sentence should not be construed as an endorsement; rather, it represents resigned acceptance of the inevitable, because, rightly or wrongly, Word appears to have become the de facto standard of the publishing industry—despite ongoing problems.

The biggest of those problems is that each new release of Word has been cluttered with additional functions, including many that most writers would never need, and the latest release is so laden with obscure and undocumented functions that it is, at best, extremely difficult to use. Sure, you might find what now passes for documentation—a brief video on the internet—but even if the procedure recommended in the video actually works, which it rarely does, you'll still have to write down each step shown in the video unless you have a photographic memory. It also definitely doesn't help that the latest version of Word is incompatible with all the earlier versions. Most readers of this book probably have an advantage over us, since you've probably used digital devices of all sorts since the cradle; we envy you that familiarity with the challenges but feel obliged to offer this little warning for those of you who may be new to Word.

So good luck, and be prepared for some acute frustration.

Another piece of equipment that can prove indispensable is a portable hard drive. Most of these are small, easy to store and use, and plug into a USB port on any computer (although even as we write this USB plugs and ports are becoming less standard by the minute). They have more storage capacity than most people could use in a lifetime, and it's a very good idea to get into the routine of storing each day's work on the portable drive before closing down the computer, then perhaps locking the drive in a file cabinet, as Steve does. Some other writers even lock their drives in a safe. Portable drives also can be taken along when you travel. That way, if something goes wrong with the files on your computer, you'll always have an up-to-date backup handy.

Why go to that trouble (though it really isn't trouble) when you could easily transmit each day's work to the supposedly unlimited storage available in the "Cloud"? Well, maybe it's an old-fashioned or pessimistic viewpoint, but what happens if the internet suddenly becomes unavailable to transmit files, as has sometimes occurred? Or if terrorists bomb the data farm that houses the

"Cloud"? Unlikely, you say—but so was 9/11. It's always safer for writers to take responsibility for the security and safekeeping of their own computer files; don't trust them to someone else.

But Wait, There's More

For sure you'll need a combination printer/copier/scanner and the software to drive it. Thank goodness those things are a lot less expensive and a lot more reliable than they used to be.

You also will want to route your incoming landline telephone or cell phone calls to voicemail answering systems. Nothing blows up a good writing session faster than having to stop in the middle of a sentence to answer an unwanted phone call.

This list of things you'll need may appear much longer than you expected, but if you're really serious about writing fly-fishing articles or books, or anything else, these are essential tools of the trade. You don't have to acquire them all at once, but over time you'll find a need for all of them, and probably more.

Once you've settled on an appropriate space to write and assembled the minimum necessary equipment, you're ready to go to work. To get started it's a good idea to establish a daily routine. If you're a morning person, at your best that time of day, then you should write in the morning, or vice versa if you're a night owl. Stick to the routine as closely as possible. You'll also discover you need to be feeling good, well-rested and free of worry or distraction to write effectively. Even if all those attributes are present, you'll still have occasional days when you just don't seem to have what it takes. It's like being a baseball pitcher: some days you just don't have your best stuff, and nobody knows why.

When that happens, don't try to force it; go fishing instead!

OK, this has been pretty easy up to now, but cinch up your waders; we're about to enter deeper water.

Make clear and simple rules, permit few exceptions to the rules and rely heavily on the chosen dictionary as the arbiter of conflicts.

—*The Associated Press Stylebook*

CHAPTER FOUR

Writing in Style

LET'S START THIS CHAPTER WITH A LITTLE EXERCISE. See if you can figure out what's wrong with the following paragraph:

"If you tie your own tapered monofilament leaders for fly fishing, try this simple formula: The butt section should be three feet of thirty-pound test monofilament (diameter .022 inch). Attach this with a blood knot to 2½ feet of 12.5-kilogram-test monofilament (diameter 0.50 mm). This in turn should be attached with another blood knot to twenty-four inches of 20-lb.-test monofilament (diameter .018 inches). Then comes another blood knot and 30.48 centimeters of monofilament with a breaking strength of fifteen pounds (diameter fifteen hundredths of an inch). Attach this with another blood knot to twelve inches of 4.5-KG-test monofilament (diameter .030 centimeters), then add twelve inches of 3X monofilament (diameter .010 inch), and another 12 inches of six-pound-test leader (diameter .007 inch), and finally one more blood knot connected to a 45.72-centimeter length of 5X four-and-three-quarter-pound test (diameter 1.524 mm), or the same length of 6X 3½-pound-breaking-strength monofilament (.004-inch diameter) as a tippet, and you're in business!"

Well, that was easy, wasn't it? The paragraph is obviously totally inconsistent in its descriptions and measurements of lengths, diameters and breaking strengths, and it would take forever to try tying a leader using those instructions. (By the way, the "editor" program in the latest version of Microsoft Word scored that paragraph as 100 percent accurate, which is further evidence you shouldn't place very much reliance on digital help.)

What's the point of the exercise? It's to demonstrate the need for consistency in writing. There's an old saying that you should never change horses in the middle of a stream; neither should you change nomenclature in the middle of a paragraph. Otherwise, we'd never be able to understand each other.

Consistency and Style

The term "style" has all sorts of meanings and applications, from clothing fashion trends to sign painting to architecture, automobile design to jazz guitar. It has similarly diverse meanings in writing, from comparing the distinctly different (though equally admired) literary styles of, say, Hemingway and Faulkner, to the more mundane but no less important practical matters of style that determine whether readers understand what we're trying to tell them.

For the purposes of this discussion, the word "style" means a set of rules or standards governing intelligent and consistent use of the language, including spelling.

Until nearly the end of the 18th century there were few such rules. Spelling was a free-for-all in which people used phonetic spellings that often led to confusion and misunderstandings, and remain one of the things that most interfere with our reading of the earliest fishing books. It wasn't until modern dictionaries were compiled and popularized that rules for spelling took hold. So dictionaries might be said to be the first stylebooks.

Since then, style has been expanded to include nearly everything else in the language—grammar, punctuation, capitalization, abbreviation, titles, numbers, etc. The point of all this is to maintain consistency in use of the language and make things easier for both writers and readers. If you're writing, and you know the rules, then you don't have to pause and wonder whether you should spell out the word "percent" or use the % symbol. If you're a conscientious reader you don't want to encounter the word spelled out in one place and the symbol in another. You want the language to be consistent. If you don't yet understand *why* you should want that consistency, it's all the more important that you keep reading.

It is impossible to overemphasize that your probable first audience—an editor—will likely be your most demanding. As a writer, you may think of yourself as a freestyle or even style-free writer who is above all those tedious technicalities, spontaneously bestowing your deepest, innermost wisdoms on a grateful public. But you must never forget that the editors to whom you hope to submit your manuscripts feel strongly about the grown-up writer's rules. If you can't use the language clearly and efficiently, editors are not going to make it through your first paragraph.

Still, there is a lot of looseness in the very concept of what constitutes good writing. From his grade-school days, Paul was complimented on his writing. This was important to him in part because he always got lousy marks for penmanship so he needed to make sure he at least had said something worth reading, in part because he idly considered being a writer many years before he had any idea of what that entailed, and in part because it started him on a long and still not entirely satisfactory search for a concise definition of good writing. When he finally did begin to publish, in the mid-1970s, he couldn't have been more grateful for the good advice about style he received from more experienced writers and editors. Much of this advice was about the countless little specifics that separate the amateur from the professional (grad-school professor: "it's *as* I say, Paul, not *like* I say"). The problem for most of us is that we have to learn all these things on the street, as it were, because, sad to say, few of our teachers were attuned to that level of using the language. But learn them we must. It may be a cliché to say you must learn the rules before you can wisely break them, but it's how writing works.

For his part, Steve started writing stories shortly after he learned to read. He also got compliments for his writing, but sometimes it got him in trouble. One teacher wrote on his report card: "The things Steve writes in his spelling book are always very funny but not always correct." More positively, another teacher predicted he would become an editor or novelist. She was right on the first count and not too far off on the second; he's now published two books of fiction in the form of short stories.

Another good way to learn how to write well is by example. Some of the best advice Paul could find back then was about what to read. As one writer friend put it, "read anything by John McPhee or E. B. White, and anything published in *The New Yorker* whether you agree with the writer or not." The best *way* to learn, however, is by getting a good grip on the rules while at the same time reading things by people who are masters of those rules and know how to use them and when to break them.

But don't get your hopes up that there is anything approaching universal agreement on these rules. There is often a bit of subjectivity and personal preference involved in deciding what's "right," but there are also practical concerns that may affect the rules depending upon the nature and form of the publication in question.

Most large newspapers have committees that make the style rules for their publications. These committees ponder such weighty questions as whether to use "Ave." or "Av." as an abbreviation for the word "avenue," or how to spell local place-names. *The New York Times Manual of Style and Usage* and *The Associated Press Stylebook* are the two most influential stylebooks in the newspaper industry, and have served Steve well during his career in newspapers, magazines, and books. The *Chicago Manual of Style* is another influential manual, often used by book publishers, especially for scholarly books. (Note that the rules of the *Chicago Manual* require more typing and usually produce a longer manuscript than any other style, which—pardon our cynicism—might be useful if you need a little padding to reach a specified number of words; on the other hand, it could force you to trim an article or book if it exceeds a specified word limit.) More as a matter of professional experience than any conscious preference, Paul still leans toward "*Chicago*," as it's known, even though he does indeed find it a cumbersome big thing. He's used it for decades for scholarly work (both humanities and science, though for science he sometimes has had to abide by the style sheet of the *Journal of Wildlife Management*) as well as for less formal writing, and will probably always stick with it from force of habit. All the answers are there even if they take a little digging.

There is no end of helpful books on writing, and many good writers get both instruction and recreation by reading them. A lot of them are pretty specialized, such as books on the mechanics of fiction: plotting, character development, dialogue and so forth. There are also many how-to books, such as "how to write for magazines," "how to sell your first novel," and other such tiresome titles. If those appeal to you and seem useful, have at them, but the beginning writer interested in understanding matters of style is probably best off starting with some of the classics. Among the most often recommended are William Strunk and E. B. White's little masterpiece *The Elements of Style* (1999), which has gone through many updates during the past century or so (see the bibliography for more on which editions are best), and William Zinsser's equally admired *On Writing Well* (2012), a longer but not overwhelming survey of many essential topics, which helpfully uses extended quotes of published works from many authors to illustrate its points. These two books are among the few that rise to the level of literary works themselves. They merit frequent rereading both to refresh the lessons and just to enjoy good, clear, lively writing.

Rules to Write By

The rest of this chapter is a "mini-stylebook" compiled from several different sources, greatly condensed and intended to address only those style questions you are most likely to encounter, or which have most often arisen in our work as writers and editors. It is by no means all you need, but it's a good start and can save you a lot of bother.

You'll see that we often emphasize brevity, which editors once considered, and still should consider, next to godliness. Many editors will urge or insist that you tighten up your prose, but you can publish for a long time without an editor ever telling you to stretch something out a bit. Admonitions to brevity are legion, such as the famous apology, "Sorry my letter is so long; I didn't have time to do it right," or the old saw that "Writing *is* revising." Strunk and White were forcefully and repeatedly emphatic: "Omit needless words!"

The style given here will save you both time and typing. Remember, though, this will give you just a small taste of what a real stylebook is like, and anyone who plans to do a lot of writing should try to acquire one as soon as possible. Meanwhile, try getting in the habit of consulting this condensed version frequently; if you do that often enough, then the rules eventually will become second nature.

Keep in mind that many specialist magazines develop their own style sheets. They have no choice, really, considering how quirky the specialist's terminology can get—fly tier or fly tyer? fly fisher or fly-fisher? There's no harm in asking a magazine's editors for their style sheet before submitting anything to them (after checking to see if it's actually offered on their website). They will likely appreciate the professionalism implied in your concern, as well as the work you might be saving them by writing your manuscript their way.

These general rules of style are presented in alphabetical order. They are followed by several helpful lists: (1) the correct spelling of commonly misspelled words, (2) the correct usage of commonly misused words, (3) a list of clichés and other terms to avoid, and (4) a section called "Rules of the Road" on how to write safely.

Some Rules of Style

Abbreviations: Do not use abbreviations a reader would not quickly recognize. Abbreviations such as FBI, CIA or NAACP are acceptable; those such as SFRB or LOWV are not. Spell out the full name on first mention. For example: Fly Fishers International on first mention, FFI acceptable on subsequent mentions. If you must abbreviate state names, use the U.S. Postal Service's method: WA for Washington, OR for Oregon, etc.

Apostrophes: Use these guidelines:

- They are used when letters are omitted from a word, or whenever a contraction is used. Examples: Isn't, aren't, I've, it's, fo'c's'le.
- They are used when figures are omitted. Examples: The gold rush of '49, the stock market collapse of '29, etc.
- Apostrophes are also used with plural forms of a single letter. Example: The coach drilled the team on X's and O's.

See also the **possessives'** entry.

At: Never end a sentence with the word "at;" use "are" instead. Most such uses of "at" are redundant anyway. Wrong: "This is where we're at." Right: "This is where we are."

Attributing quotes: Attribution for a direct quote should always be after the very first phrase or sentence of the quote. Correct: "A dam will ruin this river," said Fred Feathertwister. "It's too precious to be lost." Wrong: "A dam will ruin this river. It's too precious to be lost," said Fred Feathertwister, Conservation Trust president.

Once attribution is established, it is not necessary to reestablish it in the same paragraph. Example: "The Conservation Trust will save this river," Feathertwister said. "It's too precious to be lost," he added. If, however, you wish to quote the same source again several paragraphs further along in the story, you need to remind the reader who you're quoting, though you don't need to give the person's entire name again. Example: "We may have to have a separate fund-raising campaign to save this river," said Feathertwister, the Conservation Trust president.

Burglary, robbery: A burglary is when someone enters your home or office while you're away and steals something; a robbery is when someone uses threat or force to take something from you in person.

But: Many writers fall into the bad habit of beginning sentences with the word "But" followed by a comma. Some grammarians argue that one should never begin a sentence with the word "But," though it's hard to avoid doing so. However, starting a sentence with the word "But" followed by a comma is an even greater sin. *Do not use a comma after the word "But" at the beginning of a sentence or anywhere else the word "but" occurs.* The only exception would be in a construction like the following: "But," she said, "there should never be a comma after the word 'but.'" In that case, the comma is used to set off a clause used to attribute a quotation.

There are many alternatives to "but" (although the meanings of some are not precisely the same). They include however, although (or though), except, yet, on the contrary, on the other hand, while, despite, nevertheless, notwithstanding, regardless of (but *never* irregardless), for all that, even so.

Capitalization: Always be conservative in using capital (uppercase) letters. Use a capital only if justified by one of the following cases:

- Capitalize full abbreviations or acronyms in every case.
- Proper nouns: Capitalize nouns identifying a specific person, place or thing.
- Proper names: Capitalize common nouns, such as *south*, *river* and *street*, when they are part of the full name of a person, place, or thing. Examples: South Fork, Deschutes River, Easy Street. Use lowercase when the same words are used in subsequent references, such as *river* and *street.* Exception: In the case of animals, birds or fish, if the creature's name includes a person's name, capitalize that name but list the critter's genus or species in lowercase. Examples: Douglas squirrel, Steller's jay. Use lowercase when the name of the animal or bird uses only common words. Examples: varied thrush, gray squirrel, brook trout.
- Short titles: Capitalize short formal titles *only* when they are used immediately *before* a name. Examples: Doctor Bones, Reverend Brimstone, Representative Blowhard. (But remember that only short titles are used *before* a person's name; see *Titles*, below.)

Collective nouns: Nouns describing a group or unit, such as board of directors, team, family, crowd, audience, etc., require singular verbs and pronouns. Examples: "The board is meeting to adopt its budget." Right: "The jury reached its verdict." Wrong: "The jury reached their verdict."

Colon: Colons are most often used at the end of a sentence to introduce a list or series of items. The first word following a colon should be capitalized if it is the beginning of a complete sentence, or if it is a proper name or noun. Examples: "Building a bamboo rod requires the following: time, patience and care." "The directors of the fishing club added three measures to the annual budget: programs, awards and summer youth camp."

Colons also can be used to give emphasis. Example: "He was even-tempered: always angry."

Colons also should be used in listings such as the time of day (1:30 p.m.) or for biblical and legal citations (John 3:14, etc.). They are also used in question-and-answer sequences: Q: Did you mean to hurt him? A: Yes I did.

Colons always go *outside* quotation marks unless they are part of the quotation.

Commas: Commas should be used sparingly; otherwise they are like bricks left in the living room: People will trip over them. Some general rules:

- Before and after attribution: A comma should be used instead of a period at the end of a quote followed by attribution, and again after the attribution if the quote is continued. Examples: "It was a big success," she said. "We had many people there," the casting instructor said, "and the event was a great success."
- In a series: Use commas to separate items in a series, but don't use one before the conjunction in a simple series. Examples: "The fly pattern is red, white and blue." "He would choose any Tom, Dick or Harry."
- Introductory clauses and phrases: Use a comma to separate an introductory phrase or clause from a main clause. Example: "After losing his last fly, he had to quit fishing and go tie more." But if the introductory

clause is short, the comma may be omitted if its absence would cause no ambiguity. Example: "After nightfall she could hear salmon jumping in the darkness."

- Separating similar or identical words: Example: "What the problem is, is not clear."
- To establish cadence: Commas also may be used to help establish a rhythm or cadence in writing. In this case, they serve the same purpose as a rest mark in a musical score. Example: "The wind carries the scent of far places, of fish and kelp, of salt and distant rain."

Remember: Commas *are always* placed *inside* quotation marks.

Courtesy titles: Titles such as Miss, Ms. Mr. or Mrs. should be used sparingly, if at all, and *never* on first mention. It's best to identify a person by his or her full name on first mention, then refer to the person simply by last name on subsequent references.

Dashes: Use dashes (often referred to as "long dashes") to set off a sudden change in thought in a sentence, or signal a dramatic pause. In such cases, a dash should be placed at the beginning of the phrase and followed by another at the end.

Dashes also may be used to set off a phrase in a series of words that would ordinarily be set off by commas, as in the following example: "The land contains numerous features—old-growth forest, wetlands, wildlife habitat, scenic views—that make it desirable."

Also use a dash before an author's name at the end of a quotation. Example: "To be or not to be."—Shakespeare. A second dash is not required in such a construction.

Dates: Always use numerals without adding nd, rd, st, or th. Example: April 3 (not April 3rd or April third). You might encounter other styles for these dates, such as "3 April," but usually only in more technical publications or publications from the UK; the publication's style sheet will alert you to those.

Days of the week: Always capitalize and spell out in full, except in tabular format when abbreviations may be used (Mon., Tues., etc.).

Directions: Use lowercase for east, north, south, west, northeast, northwest, etc. Capitalize such words when they designate places or regions, such as North Fork, South Arm, the West, etc.

Ellipsis (. . .): An ellipsis is used to indicate one or more words have been deleted from a quote. For example, if the original quote was "Language is a serious matter, something to be respected," and you wanted to delete several words to condense the quote, the correct form would be: "Language is . . . something to be respected." However, if the deletion is at the end of a sentence, then the quote should end in a period *followed* by an ellipsis and then the quotations marks. For example, if the quote is "God could have made a better berry, but doubtless God never did," and you wished to delete the last phrase, the correct form would be: "Doubtless God could have made a better berry. . . ."

Ellipses also may be used to indicate a pause or hesitation in speech, or to indicate a sentence or thought that the speaker does not complete.

Etc. (et cetera): Acceptable abbreviation at the end of a sentence. Means "and others" or "and so forth." Don't be surprised, though, if an editor rejects "etc." or changes it to "and so forth" or some other fuller form. Most editors, however, would probably do the reverse; why use three words when one would do?

Exclamation point: Use to denote emphatic expressions. Avoid overuse, however, as this dilutes the impact. The more often you use it in a story, the less effective each successive use becomes. In fact, some of the best writers will tell you that you should be able to achieve the effect of an exclamation point simply by writing the sentence in such a way that it clearly indicates the special emphasis, and therefore if you are doing your job as a writer you should never need an exclamation point.

Fly tier or fly tyer: Which is correct? "Tier" is used most often but is easily confused with the identically spelled word that means a row of structures or layers. "Tyer" is not likely to be confused with anything else. Use your own best judgment and see how it goes. "Tyer" has taken over in many quarters; Paul thinks it looks silly and won't use it, but will grudgingly tolerate it if an editor "corrects" his use of "tier" and changes it to "tyer." Like so many other style rules, including quite a few given here, it's not worth an argument. Think about it tactically: it's one of those expendable personal preferences that can be temporarily abandoned if doing so keeps the editor happy. The more things like that, over which you can let the editor have his or her way, the better your chances of getting your way with the ones that matter most.

Fractions: Spell out single-digit amounts, using hyphens between words. Examples: two-thirds, three-eighths, etc. Use figures when any part of the fraction has double digits. Examples: 5/16, 11/32, etc.

Hyphens: Hyphens are used to form a single idea from two or more words. When a compound modifier (several words expressing a single concept) precedes a noun, hyphens should be used to link the words in the compound modifier. Examples: "A third-quarter deficit," "blue-green algae," "fly-fishing-only regulation" (but note that the words "fly fishing" when they stand alone and are not used as modifiers should *never* be hyphenated).

Exceptions: Do *not* use hyphens after the adverb *very* or after any adverb that ends with the letters *ly*. Right: "He was very hard-nosed about it," "He gave me a dozen crudely tied streamers." Wrong: "He was very-hard-nosed about it," "He gave me a dozen crudely-tied streamers."

Italics: Italicize titles of books, periodicals, movies, plays, television programs, operas, foreign words and phrases, and the scientific names of fish species, such as *Salmo salar*, *Oncorhynchus mykiss*, etc. Also use italics sparingly to denote emphasis.

Money: Spell out dollars and cents in text. Example: "We just don't have the dollars to buy the property." But always use the dollar sign and numerals to denote specific sums. Example: "The fishing club raised $650,000 to buy the property." For amounts of more than $1 million, use the dollar sign and numerals up to two decimal places. Example: "The asking price was $1.36 million."

Numbers: Spell out single-digit numbers (one through nine). Use numerals for double-digit numbers (10 and above). Exceptions: Use numerals in *all* cases when the number is being used as part of a compound modifier. Examples: 8-year-old boy, 180-pound man, 38-acre lake, etc. Also use numerals in *all* cases expressed as percentages: 8 percent, 36 percent, etc. Also use numerals to express page numbers in citations, vote totals, and ratios.

Try to avoid beginning a sentence with a number. When/if it is unavoidable, spell out the number, no matter what it is. Example: "Seventeen seventy six is the year the United States declared independence" (a much better construction would be: "The United States declared independence in 1776").

Also see the **fractions** entry.

Parenthesis: Place a period outside a closing parenthesis if the material inside is not a sentence (such as this fragment).

(An independent parenthetical sentence such as this one requires a capitalized first word, and takes a period before the closing parenthesis.)

Passive Voice: Once more commonly used than now, passive voice is one of those near-sins that hidebound grammarians love to hate. We mostly agree, but our hides aren't absolutely bound against the occasional use of passive voice. Just keep in mind that experience shows that most of the time a sentence written in active voice will read more easily and have a bit more rhetorical force than in passive voice. Example: "Our trout were all caught on a beautiful chicken-gut pattern." Better: "We caught all our trout on a beautiful chicken-gut pattern." If the difference seems trivial to you, trust us that it is much less likely to seem so to an editor.

Percentages: Do not use the % symbol, except in tabular matter. Otherwise spell out the words percent and percentage. Always use numbers to express percentages: 9 percent, 84 percent, etc.

Plural pronouns: Do not link singular and plural pronouns. Remember that *everyone* is singular and *everybody* is plural, so the former should be linked only with singular pronouns and the latter only with plural pronouns. Wrong: "Everyone is entitled to their opinion." Right: "Everybody is entitled to their opinion." The error is most often made when a writer is trying to avoid using a gender-specific pronoun, as in "Everyone is entitled to their opinion" instead of the awkward but accurate "Everyone is entitled to his or her opinion."

There is more to know—and thus more to adapt to in your writing—in the realm of both plural and singular pronouns today than ever before. Society, like language, evolves unevenly. You might or might not be familiar with these changes, so we should at least make sure you know you can no longer count on the usual "he" and "she" to serve all your purposes. As traditional concepts of gender have been reconsidered in light of more nuanced and broadly sympathetic approaches to personal identity, you're likely to encounter pronoun usage outside the realm of your education or reading. Many transgender individuals now choose their own preferred personal pronouns, for example replacing "she" and "her" with "they" and "them." You'll want to keep up on this, and you can start by Googling "gender identity terms" or a similar phrase.

Plurals: The rules here are many and complex, but the following guidelines will serve in most instances:

- Most words: Add an *s*. Examples: trees, valleys, lands.
- Plural words that end with *ch*, *s*, *sh*, *ss*, *x* or *z*: Add *es*. Examples: leeches, excesses, splashes, glasses, boxes, fishes.
- Plural words that end in *is*: Change *is* to *es*. Example: thesis should become theses.
- Words that end in *y*: Usually just add an *s*, but if a consonant appears before the *y*, or the *y* is preceded by the letters *qu*, then change the *y* to *i* and add *es*. Examples: fly becomes flies, jetty becomes jetties, etc.
- Words that end in *o*: If a consonant appears before the *o*, most plurals require adding *es*. Examples: grottoes, tomatoes, echoes, etc. However, there are some exceptions, such as the word "photos." Check a dictionary or your friendly spelling checker to ascertain the correct form.
- Words ending in *f*: Change *f* to *v* and add *es*. Example: Leaf becomes leaves, etc.

Possessives: For words denoting ownership, follow these directions:

- Plural nouns or names that end in s: add an apostrophe. Examples: States' rights, Bill Gates' foundation, the hostess' invitation, etc.

- Nouns that do not change between singular and plural: These should be treated the same as plurals, even if the meaning is singular. Examples: The two deer's tracks, the lone moose's antlers.
- Pronouns: The possessive forms of pronouns—mine, ours, your, yours, his, hers, its, theirs, whose—do *not* take an apostrophe. Do not confuse these forms with pronouns used in contractions such as you're (for you are), it's (for it is), there's (for there is), who's (for who is).
- Singular nouns that do not end in *s*: Add an apostrophe and *s*. Examples: The fishing club's budget, the foundation's contribution.
- Joint possession: Use an apostrophe after the last word to denote joint ownership. Example: Jack and Jill's books. But if the ownership is individual, simply use an apostrophe. Example: Jill's books.

  Exceptions: Do not add an apostrophe to a word ending in *s* when it is used primarily in a descriptive sense. Examples: citizens band radio, Authors Guild.

  However, an apostrophe and an *s* are required when a term involves a plural word that does not end in *s*. Examples: A children's hospital, a people's republic, etc. Some organizations or entities with descriptive words in their names include an apostrophe and some do not. In such cases, adhere to the entity's usage. Examples: *The Writer's Digest*, the Authors Guild, The Anglers' Club of New York.

See also the entry for **apostrophes**.

Quotation marks: Follow these guidelines:

- Direct quotations: Double quotation marks (") should be used at the beginning and end of any direct quotations. Example: "We're going to catch some fish here," she said. When a direct quotation is interrupted for purposes of attribution, either by a phrase or by division into separate sentences, double quotation marks should be used at the beginning and end of each separate part of the quotation. Examples: "We're going to catch some fish here," the guide said, "even if it takes us all day." Or: "We're going to catch some fish here," the guide said. "Even if it takes us all day."

- Quotations within a quotation: Use double quotation marks to begin the quotation, single quotation marks (') to begin and end a quote within a quote, then double quotation marks again at the end of the quote. Example: "I think when Izaak Walton said, 'A man can't lose what he never had,' he'd just lost a fish." A single quotation mark followed by double quotation marks should be used when the sentence ends with a quotation within a quote. Example: "I remember Izaak Walton said, 'A man can't lose what he never had.'"
- Placement with other punctuation: Periods and commas *always* go *within* quotation marks. Dashes, semicolons, question marks and exclamation points go within quotation marks when they apply to the quoted matter only. They go outside when they apply to the whole sentence.

Said: When you attribute a quotation or statement to any person, *always* use the word *said.* Do not feel like you must vary this by using such words as commented, stated, declared, pointed out, insisted, asserted, etc. Use of the latter words is a sure sign of an inexperienced writer who overestimates the importance of the attribution. Veteran novelists and other writers of extended dialogue know that readers are so accustomed to "said" that we essentially read right over it, no matter how often it's used in a long back-and-forth conversation. The great virtues of the word *said* are that it is short and absolutely neutral. Nobody will ever derive any unintended nuances from the word "said," which makes it the safest of all words to use for attributing quotes. Once you have established a consistent back-and-forth exchange between two people in a conversation, you can sometimes just write the quotes and drop the "Bob said," but don't do that for long or readers will lose track of who is speaking.

Semicolons: Use a semicolon to alert the reader to a more significant change in thought or information than would be indicated by using a comma, but a less significant change than would be signaled by inserting a period followed by the beginning of a new sentence. Note, however, that some grammarians contend semicolons should not be used at all to signal a change of thought or information within a sentence; they say it's better to use a period and begin a new sentence. Use your own judgment.

Semicolons also are used to help clarify elements of a series. Example: "Survivors include a son, John Troutseeker of Bozeman, Montana; daughters Jane Cane-Rodd of Eureka, California; June Hookkeeper of Jonesboro, Georgia; Janet Rollcast of Forks, Washington; and a sister, Janice Backcast of Boulder, Colorado."

Semicolons also may be used to link independent clauses when a coordinating conjunction, such as *and*, *but* or *for*, is omitted. Example: "The rod was shipped two weeks ago; it was finally delivered today."

Split infinitives: Avoid splitting infinitive forms of a verb, such as *to run*, *to assist*, etc. Right: "To go boldly where no one has gone before." Wrong: "To boldly go where no one has gone before." Captain Kirk wouldn't approve, but there it is.

Time: Use a.m. or p.m. to denote time of day. Always express the time in numerals. Examples: 4 p.m., 11 a.m. (Note this is an exception to the rule that single-digit numbers should be spelled out.) Also acceptable: 4 o'clock in the afternoon, 11 o'clock in the morning. *Never* say 8 p.m. tonight or 9 a.m. this morning; both are redundant. If you wish to refer to the time of something happening on the day you are writing or publishing a statement, say 8 p.m. today or 9 a.m. today.

Titles: When identifying titled individuals or attributing quotes to such individuals, follow this rule: *Short* titles (one or two words) go *before* the person's name and are capitalized; *long* titles (more than two words) go *after* the person's name and are *not* capitalized, except for proper names. Examples of the short-title form: President Joe Biden, State Senator Margaret Filibuster. Examples of the long-title form: Millicent Muckraker, assistant senior managing editor; Lance Longcaster, executive assistant deputy casting instructor.

In general, do not abbreviate titles except for courtesy titles (Mr., Mrs., Ms., etc.)

Frequently Misspelled Words

Accommodate (always two c's and two m's)

Acknowledgment (no e between the letters g and m)

Harassment (only one r)

Judgment (only one e)

Led (past tense of the verb "lead")

Likable (only one e)

Receipt (e before i in this case)

Separate (there's always "a rat" in "separate")

Vacuum (two u's)

Frequently Misused Words

Affect, effect—As a verb, *affect* means to influence. Example: "Rain will affect the rate of erosion." Do not confuse with the noun "effect," which means result. Example: "The effect was overwhelming.

All right—Two words. Never "alright."

A lot—Two words, as in "there are certainly a lot of fishermen." Many people merge it into one word, which is incorrect (perhaps they confuse it with the word "allot," which has an entirely different meaning).

Continual, continuous—Continual means a steady repetition, over and over again. Example: "Negotiations over the land purchase have been continual." Continuous means steady, uninterrupted, unbroken. Example: "The view was of continuous forest."

Council, counsel—The former is a deliberative body. The latter is someone who advises. Don't get them confused.

Ecology, environment—Ecology is the study of the relationship between organisms and their surroundings. It is not synonymous with environment, which refers merely to surroundings. This is especially a problem for writers who are discussing some *environmental* issue, such as air pollution, but refer to it as an *ecological* issue. It could indeed also involve ecological factors—most environmental issues do—but the greater context is the environmental one.

Either or, neither nor—And don't you forget it!

Farther, further—Farther is a measurement of *distance*; further is a measure of *degree.* Correct: "It's farther from Boise to West Yellowstone than it is from Bozeman." "Let's not push the matter any further."

Flounder, founder—A flounder is a fish. The verb form *to flounder* means to flop around. The word *founder* means to bog down or sink. Example: "The ship floundered in heavy seas, then foundered."

Forego, forgo—To *forego* means to go before. To *forgo* means to abstain from.

Forward, foreword—Forward is a direction. Foreword is the introduction to your book. But lots of amateur writers list it as a forward, which is a sure sign to an editor that the writer doesn't know what he or she is doing.

Its and it's—Probably the second-most misused words in the English language. The apostrophe is used *only* when the word is a contraction of the words "it is." There is no apostrophe in the possessive form of the word "its." Wrong: "Its likely Congress will adopt the budget." Right: "It's likely Congress will adopt the budget." Wrong: "The council adopted it's budget." Right: "The council adopted its budget." (See the **apostrophe** and **possessives** entries in the Style Guidelines).

Last—Not a synonym for *past*. To say "it happened in the last week" indicates there will never be another week. The correct form is to say "it happened in the past week."

Myriad—This word is tricky. Stylebooks vary on its use. You'll see it used in a variety of ways in print, even in respectable, high-tone publications. In short, "myriad" is not the equivalent of "lots." You may say "lots of," but you should not say "a myriad of." Wrong: "A myriad of waterfowl were flying over the river." Right: "Myriad waterfowl were flying over the river"

Off of—The "of" is unnecessary. It's enough just to say "off."

Only—Misplaced 99 percent of the time. Nearly everyone puts it near the beginning of a sentence instead of as close as possible to the word it modifies, where it belongs. Doing that also usually changes the literal meaning of the sentence. For example, consider those annoying insurance company TV commercials that say, "only pay for what you need." The literal meaning of that sentence is exactly the opposite of what the insurance company intended to say; it means *you* are the *only* person who can pay for what you need; nobody else can. What they meant to say was "pay for only what you need," which means that everyone can pay for only what he or she needs. It's amazing to think of the millions of dollars wasted on commercials saying the wrong thing, just because a word was misplaced. Remember: "only" always belongs as close as possible to the word or words it modifies.

Over—Frequently misused as a synonym for "more than." When comparing numbers, it's always more precise to say "more than." Wrong: "The fishing club has over a hundred members." Right: "The fishing club has more than a hundred members."

Plan on, plan to—In our lazy speech patterns, it's common to say "plan on." But you should never say that in writing. Always say "plan to."

Presently—Does not mean "now." "Currently" means now. "Presently" means soon. But if you mean now, why not just say now? Maybe you don't even need the "now."

That, which—There is a simple rule for determining when to use "that" and when to use "which." The rule: If the modifying phrase can be embraced in commas, use "which"; if it can't, use "that." Examples: "The document, which is on the table, needs signatures." Or: "The document that is on the table needs signatures."

Try and, try to—Here's another lazy speech pattern. It's common to say "try and" in speech but avoid it in writing. Right: "I'll try to catch a fish" Wrong: "He was going to try and see if he could catch a fish."

Unique—Means one of a kind, there's nothing else like it in the world. So nothing can be most unique, very unique, nearly unique, etc. It's either unique or it isn't. Don't use adjectives or adverbs to try to modify or qualify this word; it can't be modified or qualified.

What—Often used unnecessarily. Wrong: "The weather is better than what we saw yesterday." Right: "The weather is better than we saw yesterday." Or, to improve it even further, "The weather is better than it was yesterday.

Clichés and Other Terms to Avoid

Clichés are words or phrases that are used far too often. They are unmistakably the work of lazy or unimaginative authors. Take the time (which usually isn't much) to think of a good substitute.

Fishing writing has always had these—screaming reels, tight lines, gin-clear water, blizzard-like hatches, regattas of mayfly duns, gossamer tippets . . . on and on. For a while in bass fishing circles a crudely vivid favorite involved going out to "gouge some sows." It's sad in a way, because so many of these were quite evocative the first time they were used, and were even serviceable for a few uses after that. But quickly their use became almost ritualized; an old book was invariably "classic," as were an increasing number of other things there for a while.

Here's a list of some of some common clichés to avoid, lest you annoy readers:

At the end of the day—If we're lucky, we'll live to see the end of this worn-out phrase.

At this point in time—It's not only a cliché, it's redundant. Just say "at this point" or "at this time," or even just "now."

Awesome—It's become awesomely overused.

Classical—Never use this if you mean "classic," and never use "classic" anyway.

Each and every—This cliché has been around almost since Jurassic times. It's also obviously redundant. It's OK to say either each *or* every, but *never* both.

Going forward or moving forward—For some reason, bureaucrats love this one. It's not just a hugely overused cliché, but also meaningless; what other direction could the planet or time be moving? If you must use this sort of phrase, remember that "from now on" is still functional.

Icon, iconic—These words have become huge clichés. Don't help them along.

Literally—If you're stating a fact, this word is redundant. For example, "The fish literally took me down to the backing." How else could the fish have taken you down to the backing other than literally?

New normal—Who even remembers the old normal anymore? Find another way to suggest this change in whatever context you're describing.

New record—Redundant. Every record is new; otherwise it wouldn't be a record.

Perfect storm—Has become the perfect cliché. Avoid it like the plague!

Proactive—Arrgh! Double arrgh! It's not only a horrible cliché, it's redundant. It's enough just to say "active."

Reaching out—Not a synonym for "calling; just say "calling" instead.

Remains to be seen—Superfluous. Everything remains to be seen!

Takeaway—As in "the takeaway from the meeting was that we need fewer meetings." Let's take away this annoying word, which has lately become a cliché.

Tight lines—The ultimate fishing cliché. Try to think of something more original to say.

Winter wonderland—Let's retire this tired old turkey. Use your imagination and come up with something better.

Rules of the Road

"By necessity, by proclivity, and by delight, we all quote," said Emerson. He was right; every writer has at least occasional need to quote from the work of others. But that often means using copyrighted material, so it's important to know the rules for using such material—what you can and cannot do safely. This section provides basic information on how to quote from other publications without risking copyright violations.

It's unlikely you'll ever risk libeling someone, but it's still worth knowing what constitutes potential libel so you can avoid it. This section also provides basic information about libel.

But let's start with copyrights:

What is a copyright? A copyright is the equivalent of legal title to intellectual property. It vests the author of a literary work (published *or* unpublished) with all the rights to that work. With few exceptions, that means no one else can lawfully quote from, excerpt, or reproduce any part of the work in any form without the author's express permission. Under current law, the term of a copyright extends for the lifetime of the author plus 70 years. The author also may transfer the copyright to another person or entity, either by assignment or bequest.

How are copyrights obtained? The U.S. Library of Congress maintains a registry of copyrights. Those who wish to claim authorship of a literary work must submit a completed registration form, at least one copy of the "best edition" of the work, and pay a fee to have the copyright registered. This must be done within five years of the time the work is completed or published. Registration provides a paper trail of ownership in case of a future copyright dispute.

How can you tell if a work is copyrighted? Current law (Title 17, U.S. Code) specifies that *any* written work, published or *un*published, is *automatically* entitled to copyright protection. This includes works that are not registered with the Library of Congress (the only difference between an unregistered copyright and one registered with the Library of Congress is that the copyright owner may face a more difficult task proving ownership if the work is not registered). It also is no longer necessary for a work to display the standard copyright notice—e.g., © Copyright, 2023, by So-and-So—to enjoy copyright protection (though it's a good idea to use it anyway).

The essential point is you must assume that *any* document you see in print is copyrighted.

What happens if you infringe on a copyright? If you quote copyrighted material without permission or without following established legal guidelines for "fair use" (see below), nothing may happen—if you're lucky. Perhaps the copyright owner will never know you violated the copyright. Or perhaps the owner will find out but decide the matter is of little consequence. Or maybe the owner will write you a nasty letter. But if the copyright owner can prove that your use of the material had a significant adverse effect on the market value of the original work, then you could be the target of a suit for damages. Such suits are extraordinarily expensive to defend. Not only that, but if you lose, the court will award damages in the amount suffered by the copyright owner and you will be liable. In addition, if the copyright owner can prove the infringement was deliberate on your part, the court may add punitive damages to the bill.

That's the worst case, unlikely to happen. But the best way to avoid such consequences—and remain on good terms with the people whose work you quote—is to respect copyrights and play by the rules.

When is it safe to use copyrighted material? Copyright law and legal precedents allow the use of copyrighted material under the following circumstances:

1. With the owner's permission. Absolutely the safest way to use copyrighted material is with the express written permission of the owner. In making application for such permission, you must specify *exactly* what portion of the copyrighted material you want to use and *exactly* how you intend to use it. Don't be surprised if the owner seeks compensation for use of the material. This is especially likely if the request is to use a substantial portion of the work.

For this reason, plus the fact that it is often impractical or time-consuming to seek written permission from a copyright owner, this method is seldom used.

2. Under the Doctrine of "Fair Use." Courts have long held, and copyright law now specifies, that there is a right of "fair use" of copyrighted material without the owner's permission and without compensation to the owner. In determining whether a use is "fair," the law applies the following tests:

What is the intended use? If the intent is to use copyrighted material for public benefit or charitable purposes, then it is more likely to be considered a "fair use."

What is the status of the user? If the entity seeking to use the material is a nonprofit organization, then it is more likely to be considered a "fair use."

What is the impact on the market value of the original work? The intended use must not significantly affect the market value of the original work. For example, unauthorized publication of the final chapter of a best-selling mystery, in which the solution to the mystery is revealed, would obviously detract from the market value of the original work and would definitely not be considered "fair use." But use of a short excerpt or quotation from a copyrighted work that does not detract from the market value of the original work—and, in fact, may even enhance its value—is likely to be considered "fair use," especially if it is done by a nonprofit organization for public benefit.

The law does not specify any particular number of words that may be used under the "Fair Use" Doctrine, but courts have generally considered 600 words as the maximum—so long as the use of that many words does not adversely affect the market value of the original work. But if the original work is only 700 words in length, then using 600 words would obviously exceed the bounds of "fair use." So the length of a work must be considered in determining "fair use."

The nature of the work must also be considered. The "Fair Use" Doctrine does *not* apply to poems, verses, or song lyrics. Because of the inherent nature of such works, *even the use of two or three words without the owner's permission could be considered a copyright violation.* So it is best to avoid all such uses, unless you have the owner's permission or the work to be quoted is in the public domain (see below).

You should apply the "fair use" tests to any copyrighted material you wish to use. If you believe the material passes these tests, and the material is to be used without the permission of the copyright owner, then *you must still credit the owner in every case.* In other words, if you use a quote from a copyrighted work, you must always cite the author's name. If the quote is more than a few words, you should also cite the title of the work from which it came.

Is there material that cannot be copyrighted? *Copyrights do not apply to ideas or facts; they apply only to the words used to express ideas or facts.* So if you use different words to express the same ideas or facts—that is, if you *paraphrase*—then you are not in violation of any copyright.

Example: If Patrick Henry's famous phrase, "Give me liberty or give me death!" was copyrighted, you could not repeat it without permission. But you could still say something like: "All things considered, if I couldn't live in a free society, well, then I suppose I would rather not live at all." It might lack the pop and crackle of old Patrick's famous phrase, but it still gets the idea across.

What is "public domain" material? Material for which all copyrights have expired is said to have entered the "public domain." It is material, published or unpublished, for which all copyrights have expired. Once such material has entered the public domain anyone is free to use it without violating any other person's property rights. This change in a piece of writing's copyright status is the reason it is so easy for many modern publishers to reprint older fishing books; they don't have to share the profits with an author.

However, we must offer two warnings about using public domain material. First, keep in mind that the legal field of intellectual property rights is vast, and grows more contentious and land-mined all the time. There are some slippery details in the interpretation and enforcement of the public-domain rule and concept. We have both encountered them, and it took considerable effort and expense to resolve them. Any time you wish to use material that appears to be in the public domain, you should work closely with your editor to make sure you're on firm legal ground. Though book contracts tend to require the author to take full responsibility for ensuring the proper acquisition of all necessary rights and permissions involved in quoting other sources, it is obviously to your publisher's advantage to make sure you get this right. In some cases you may even need to seek independent legal advice. Don't hesitate to do so; leave no room for doubt.

Second, even if there is no question the material you're interested in is in the public domain, all other traditional ethical concerns still apply. When the material enters the public domain, the only thing that has changed is that you no longer have to pay anyone to use it. You are not entitled to publish it as your own writing. Always make certain—by means of quotation marks and by fully acknowledging or citing the source—that your readers understand that you did *not* write it. Again, leave no room for doubt. If you use public domain material without giving appropriate credit it's nothing but plagiarism. Similarly, to "revise" it or do a bit of paraphrasing to suit your needs is as unethical as claiming you wrote it. Only if the material is so old as to contain archaic language, or is otherwise in need of slight "modernizing" to make it accessible to modern readers, might revision be tolerable, and even then only if you clearly alert your readers that you have made such changes. Again, it's always best to run these decisions past your editor for professional guidance. There are long-standing forms and standards for such textual modification, and your editor should be able to help you locate them.

Countless tons of government documents are also by law public domain from the moment they're published, but even these can be tricky; if, for example, such a document quotes verbatim material from another publication that *is* protected. However, that doesn't mean it still can't be useful. Remember, however, that even when public domain material is used, out of respect for the public domain document and its author, you should always, without fail, acknowledge the full source of such a quote, if the quote is more than a few words long. If you don't do these things, you're at risk of being accused of stealing someone else's work, and whether it's copyrighted or not, you're going to end up with a big black eye for your neglect of good practice.

What is plagiarism? Plagiarism is the use of another writer's work without permission or giving credit. This applies whether or not the work is copyrighted. Unless use of the plagiarized material detracts significantly from the value of the original, there are no legal penalties for plagiarism (although if the material is copyrighted, then the plagiarism may be actionable under copyright law). Nevertheless, plagiarism is generally considered a form of literary larceny. It's also a sure way to earn the enmity of the literary community, if not the general public. So it should be avoided at all costs.

Unfortunately, plagiarism is not unknown in fly-fishing magazines. Few magazines have escaped publishing articles unknowingly stolen from the work of other writers, even by someone who surely knew better. These cases often come to light when the magazine hears from the original author whose work was stolen or from readers who recognized the article as having been copied. These matters nearly always go unpublicized because the magazine that unknowingly published the plagiarized material is too embarrassed to admit it, and the original author has no remedy other than taking the matter to court, which rarely happens because the amount of money involved is too small to justify a costly lawsuit. Usually the only consequence is that the author who committed plagiarism will be forever blacklisted by the magazine that published the stolen article, and very likely by other editors whom your editor warns about your nefarious behavior.

What is libel? Libel is the publication (including on the internet) of false defamatory information that results in damage to an individual's reputation and/or business. Libel differs from slander in that the latter is the dissemination of damaging false information by oral means, as in a speech, radio broadcast or television program.

There are two general types of libel: Libel *with malice*, which means the deliberate publication of a defamatory falsehood with intent to injure, and *negligent* libel, which means the negligent publication of false defamatory information (not on purpose). Persons who suffer damage to their reputations and/or businesses as a result of such publications may seek damages from those responsible. Usually this includes not only the person who wrote the defamatory article but the organization or entity responsible for publishing it. It also may include any other outlet that published the same material. For example, if a false defamatory article appears in a newspaper and is picked up by a wire service and transmitted to other newspapers around the country, or if it appears on the internet and is picked up and used by other sites, then any of the other newspapers or sites that publish or post the article may be included as defendants in a libel suit.

Verdicts in libel suits can result in huge penalties. This is especially true in cases of libel with malice, or deliberate publication of false defamatory information with intent to damage a person's reputation or business. In such cases the verdict usually includes not only compensation for damage to the victim, but also punitive damages to punish the defendant for the deliberate publication. However, malice is difficult to prove, so verdicts of this type are relatively rare.

Cases of negligent libel are far more common. They often involve instances of mistaken identification. For example, a newspaper may publish a story that an individual has been charged with a crime without taking care to distinguish that individual from others who may have the same name. That's why it's so important in potentially libelous situations to identify persons very carefully so there is no chance for mistaken identification. Even a simple typographical error, unnoticed by a proofreader, can sometimes result in libel. A proverbial example is the account of a newspaper that published what was intended to be a laudatory story about the charitable and volunteer activities of a socially prominent woman. Among other activities, the article described her as the president of a home for unwed mothers, but when the story appeared in the paper the letter "p" had been dropped from the word "president." You can imagine the consequences.

Cases of negligent libel often can be settled out of court by publication of a retraction, correction or clarification. But the cost in embarrassment to all parties is still high. That's why it's so important to read your copy carefully and make certain it's letter-perfect before it goes to the public.

Can anyone become a victim of libel? No one can be libeled by the truth. If a defamatory statement is provably true, then the target of that statement cannot prevail in an action to recover damages for harm to his or her reputation or business. Truth is an absolute defense against libel. But private citizens can be libeled by publication of a *false* defamatory statement.

Not everyone is considered a private citizen, however. The U.S. Supreme Court has established a class of "public figures" who are generally immune from libel. "Public figures" are defined as persons who have deliberately thrust themselves into the limelight—politicians, movie stars, rock stars, etc. The legal theory behind this classification boils down to the notion that if you can't stand the heat, you'd better stay out of the kitchen, but if you *do* choose to enter the kitchen, then you have no right to complain about the heat. The theory also assumes that public figures have greater access to the media than private citizens, and therefore possess greater means of replying publicly to any false defamatory statements made about them.

Under this doctrine, you can say almost anything you want about a public figure without fear of consequences. That's how politicians get away with false "attack ads" about their opponents and supermarket tabloids almost always get away with publishing false stories about celebrities.

The definition of a public figure is not always clear-cut, however. Is a county commissioner or supervisor a public figure? The answer almost certainly is yes, since a candidate for the office would have to enter the public limelight deliberately. But what about an appointed position, such as a county parks director? The answer in that case is much less clear, and there is at least a high probability that a person occupying that position could make a strong argument that he or she is not a public figure.

In summary: Be careful what you say about anyone, especially individuals who are clearly not public figures. Check your copy. Check it again. Let others read it. Double-check the accuracy of every word before it goes to the public. Make sure you say it safely.

It's also worth noting that virtually any book publishing contract holds the author responsible for any libel that may occur from publication of his or her work and indemnifies the publisher from any loss that may occur. Another big reason to make sure you say it safely.

Final thoughts: Use your spelling checker. Then use it again. Then use it one more time. When you think you've finished your article, or whatever you're writing, let it simmer for a day or two, or even a week, then take another look at it; you'll be surprised how many things you'll see that need to be changed. After you make the changes, let it simmer again and repeat the process.

Another helpful thing is to read your story out loud, just to see how it sounds. If it sounds awkward in places it probably is, so fix the problem(s) and then try it again until it sounds right.

The time to begin writing an article is when you have finished it to your satisfaction. By that time you begin to clearly and logically perceive what it is that you really want to say.

—Mark Twain

CHAPTER FIVE

Writing for Fly-Fishing Magazines

Magazines are a good place to start because magazine stories are much shorter and easier to write than full-length books and because stories are a lot easier to get published. They average a couple of thousand words at most, often much less, while a book typically requires 60,000 to 100,000 words or more. And although magazines pay much less than books, they pay faster. Also, if you write for magazines, there's no need to worry about having to promote your work once it's published, or face criticism from reviewers when the book is published (though the magazine's readers can be pretty cantankerous, even vindictive, which is just another reason to make sure you know what you're talking about).

There are, however, a couple of downsides to writing for magazines. As we've seen, they are ephemeral; they don't last long before ending up in the recycling bin, while books are relatively permanent. Some writers care about this and some don't, but it's pleasant to think that your words might last longer than a couple of months. Also, depending on the type of story you're writing, many magazines require lots of top-notch color photos to go with a story, so unless the magazine is willing to assign a photographer to your story, or happens to have a good stock of photos that apply to your story, you need to be a good photographer as well as a good writer. You're even better off if you're a good artist and can provide quality illustrations for your piece; quite a few of America's best-known fly-fishing writers, from Louis Rhead a century ago, through Preston Jennings, to such modern writers as John Betts, Ernest Schwiebert, Lee Wulff and Dave Whitlock sometimes enhanced their writing with their own accomplished, distinctive artwork. Who doesn't recognize a Whitlock drawing?

The Evolving State of the Craft

Like pretty much everything in today's web-saturated world, the fly-fishing magazine business has changed dramatically in just the past 30 or 40 years. We recall when writers had very few outlets for their fly-fishing articles, stories, and essays. In the early 1970s, as fly fishing experienced a great surge in popularity and eventually became—for better or worse—fashionable, fly fishers were more and more identified, both by themselves and by all the people who wanted to sell them something, as a distinct segment of the greater fishing community, a segment deserving of its own specially focused magazine.

A writer's only choices then were *The Flyfisher*, the first modern fly-fishing magazine in North America, published by the Federation of Fly Fishermen (now called Fly Fishers International); *Fly Fisherman*, the first commercial fly-fishing magazine; *Trout*, published by Trout Unlimited (and not limited to fly fishing), and *The American Fly Fisher*, published by the American Museum of Fly Fishing. None of these magazines paid much, and those that served as voices of nonprofit organizations sometimes even viewed writers' contributions as donations to their good cause (even today, "payment" from *The American Fly Fisher* consists of 12 copies of the magazine in which your article is published). One reason they don't have to pay much, besides their not having the money to do so, has always been that there have been so many of us eager to write for them.

Back at the beginning of fly fishing's renaissance in the 1970s, several other magazines, including *Gray's Sporting Journal*, *Sports Illustrated*, the "big three" of *Outdoor Life*, *Field & Stream* and *Sports Afield*, and several other general-interest fishing or outdoor magazines, took occasional fly-fishing pieces, as did a number of newspapers and even some of the fancier tackle catalogs. Most of these still do, especially *Gray's*, but they're a longer shot than the magazines devoted entirely to the sport.

There were also a very few specialized magazines focusing on a specific region or species of fish, most notably the *Atlantic Salmon Journal.* Besides these choices, there were some distinguished but nonpaying outlets available to members of a few fishing clubs: The *Bulletin* of the Anglers' Club of New York, the Theodore Gordon Flyfishers' *Random Casts*, and *The Creel*, published by the Flyfishers' Club of Oregon, all excellent publications written by their members, who were some of the most prominent fly fishers of the day.

There was also a time when airline magazines had an appetite for fly-fishing stories, but usually only if they were essentially travel pieces about locations where their planes regularly landed.

Even more rarely, fly-fishing stories, almost always those written by writers who were already famous, appeared in such big-time publications as *Esquire*, *Harper's* and *Life* magazine. These unlikely venues will always be there for the lucky one-in-a-million writer who is somehow perfectly placed to contribute to them. For those lucky few, there have also been fly-fishing articles, usually quite good ones, in *Audubon*, *National Geographic*, and even *The New Yorker*. But these were such rare events as to have no bearing at all on the careers of 99 percent of fly-fishing writers.

Still, it's worth your trouble to keep an eye out for unusual opportunities. Fly fishing has oozed into all sorts of unlikely corners of the publishing world, so work to your strength. Over the years Paul, being a historian by training, has placed fly-fishing-related articles in half a dozen popular and scholarly history magazines. Steve often wrote for *Sports Illustrated* back in the days when it still considered fishing a sport.

Today, by contrast with only a few decades ago, who can even keep up with the rapidly changing landscape of fly-fishing magazines, both print and online? Now we have *American Fly Fishing*, *FlyFish Journal*, *The Drake*, *Fly Fusion*, *Tail Fly Fishing* and others, but nearly as many have recently disappeared, including *Fly-Fishing Heritage*, *Flyfishing News*, *Fly Rod & Reel*, *American Angler*, *Eastern Fly Fishing*, *Northwest Fly Fishing* and *Southwest Fly Fishing*.

However, there are now more opportunities for women fly-fishing writers than ever. This shouldn't be surprising; fly fishing has always mirrored changes in the society of the time. By a recent count there are more than 40 women's fly-fishing clubs in North America, and reflecting this increase in the numbers of women fly anglers, the pages of most fly-fishing magazines have been opened to articles for, by or about women. One magazine, *DUN*, was started mainly to serve women fly fishers, but after publishing several very promising print issues it ran into trouble during the COVID-19 pandemic and ceased publishing. Other magazines have added women to their staffs, which is a sort of good news-bad news situation because it employs more woman writers but also means the articles in those magazines are usually staff-written, not by freelancers. But as publishers continue catering to women readers, potential markets for women writers can only increase. We all should hope the sport will continue to broaden its welcome to anglers other than the white men who dominated fly fishing for the past few centuries.

These welcome changes haven't led to many changes in what magazine writers are paid, however, and although we may already seem to have excessively beaten the dead horse of how little money there is to be made, it seems important to reemphasize it. A handful of skilled, gifted, and lucky fly fishers do make a decent living writing about it, and more power to them; they've earned it, but they amount to Cinderella stories in the big picture. It's unlikely their number will increase.

Getting into Print

This chapter is intended to help you get started so you can eventually reach the point of writing saleable stories for fly-fishing magazines. But first, a few words of caution: As this is written, there is a lot of excitement about AI, or artificial intelligence, including its reported ability to generate "original" magazine articles or books on command. Our advice: Avoid it; avoid it like the plague.

The Oxford English Dictionary defines "artificial" as "made or produced to copy something natural; not real." So articles or books produced by that method are not real; they are merely computer-generated collections of bits and pieces taken from articles or books written previously by humans, whether the subject is fly fishing or anything else. For someone to create one of these abominations and put his or her own name on it, as if it were an original work, is unethical to say the very least. And if that isn't enough to keep you awake at night, there are legions of lawyers waiting eagerly to file lawsuits over copyright law violations stemming from the use of AI to create articles or books. You don't want to see your name on one of those expensive complaints.

Artificial intelligence may yet prove to have some tangible benefits, but trying to use it as a substitute for the difficult but rewarding work of original writing will never be among them. Confine your work with artificials to tying flies.

The basic principles outlined here were conceived long ago but remain valid for today's magazines, books and digital media—even some modern magazines that publish only the print equivalent of TV "sound bites" instead of real stories. If you stick to these principles, you won't go wrong.

It should be obvious we're talking here about nonfiction in all cases. Very few fly-fishing magazines publish fiction (as do very few fly-fishing book publishers). Even fewer magazines publish poetry, but if you study all their posted submission guidelines you might find one or two.

The best way to begin is by choosing a target magazine and studying it closely. Make sure it's a magazine that would be interested in what you want to write. This may seem intuitive, but you'd be surprised how many poorly targeted manuscripts there are, so we're going to start with some bad examples: If you want to write a story about how to splice fly line to backing, you shouldn't try to sell it to a magazine devoted to articles about fly-fishing destinations. As its name implies, *Fly Tyer* magazine is dedicated to fly tying, so don't approach them with a story about rod repair. Look up the target magazine on the internet and see if its submission guidelines are posted; some magazines describe their requirements at length. *The Writer's Market* also lists the requirements of hundreds of magazines, although it includes very few fly-fishing magazines.

Next, study the target magazine closely, both to get a better idea of the type of articles it publishes and the way those articles are written—their structure, approach and formality, or lack of it. If it publishes stories about fly-fishing venues or destinations, check to see if they include descriptions of the surroundings, accommodations, people, etc.

Same for details of tackle, flies and other requirements. Do they highlight funny or unusual incidents? Once you become thoroughly familiar with the magazine's publishing format, you can emulate it in your own story.

Another reason to study the magazine you want to write for is to keep from duplicating stories it has already published. An additional reason is to gain a sense of the general tone of the magazine.

It's a rare fishing article that doesn't require some background legwork: research, reading, interviewing, or all the above. Sometimes it's very little, sometimes it's a lot. In our experience, once we start to look into a subject, we almost always find there's more homework to be done than we expected.

Beyond offering a few general bits of advice, we can't help you become a careful researcher, but here are some tips: Check all your facts and get at least two sources for anything controversial. If you went to college and took some classes that required writing extended research papers, you have presumably learned the basics of gathering background information on a topic. If you took some journalism classes, even better.

The Art of the Interview

For some stories, library reading isn't enough, so don't shy away from conducting interviews. Interviewing people is its own art form, and entire books are written about proper techniques. We'll just offer a few basics here.

First, this shouldn't be necessary to say, but judging from experience it is: for Pete's sake be on time. And do your homework before you arrive for a scheduled interview. Don't walk in with absurdly general questions like, "So tell me all about trout management" or "Why catch-and-release?" In these days of the so-called 24-hour news cycle too many journalists have become accustomed to conducting interviews with essentially no prior knowledge of the subject; they're just there, as they say, to "get some quotes" to string together and flesh out their article with the voices of presumably knowledgeable people. Don't think that way; what the interviewee says is very important, and it's your job to weave his or her voice into a greater coherent narrative of your own.

Many interviewers record the entire conversation, which is now easy with most cell phones. Always ask permission to do so, and don't argue if someone says no. Just get ready to write really fast when they talk. Botched quotes, to which the interviewee objects to as not representing what was said, can get you into serious trouble with both the interviewee and the editor who published your interview.

Make sure you have some familiarity with the subject and especially with the interviewee's work. These days it's ridiculously easy to learn about most people you might find yourself interviewing, starting with what other journalism exists on your subject. If you're interviewing a public figure of any note, odds are excellent there will at least be an online profile of the person, and possibly a personal website that offers his or her views on whatever topics you want to discuss. If you're interviewing someone who publishes, whether formally (e.g, a scientist or other scholar) or informally (e.g., leader of a conservation organization who writes editorials), don't hesitate to do a Google Scholar search of your interviewee's publications and read the relevant ones. You'll be amazed what a difference it can make if your first question is about some point he or she made in an article in a recent issue of *The Journal of Fish Biology*. She'll immediately know she's working with a pro and will be much more willing to give you time and thoughtful attention.

Keep your interviews non-adversarial. Confrontational journalism is showmanship for television's political shouting matches. Your interviewee will quickly figure out whether you're there to learn or merely to argue from some already established position of your own, and will respond—and shut you off—accordingly. This is especially important if you're writing about a controversy, say in fisheries management.

None of the above is intended to set up your interviewee as being the soul of probity. Like any cross section of the public, some interview subjects will be inarticulate, some grouchy, some shy. A very few may just be nuts, but you should always try really hard not to assume that. Depending upon the subject, some might be suspicious of possibly uninformed generalists like you, and some are just hostile or uncooperative by nature. Even the most distinguished scholars invariably have their own perspective, and they may or may not be willing to call it a bias. Much worse, long-standing journalistic cynicism about interviewing politicians is sometimes sadly justified, as suggested by the old advice that the way to tell if a politician is lying to you is to watch his face very carefully; if his lips are moving, he's lying.

The reality is more subtle; almost everybody you might interview will be sincere in their convictions and their confidence in the information they impart to you. The best such sources will even point out the areas where they regard their information to be incomplete or less reliable. Your conversation with them is a diplomatic dance. Be respectful and grateful for the time they're giving you. Never forget that many of them are devoting their careers, even their lives, to a subject that you've just dropped in on from another realm. Be nice.

And never forget: This isn't about you! You're not having a normal conversation with your interviewee, in which you're exchanging opinions, swapping stories, or otherwise having a typical back-and-forth chat. Unless an interviewee clearly wants to engage in such talk, stick to your job; ask intelligent questions that will elicit thoughtful responses, and listen carefully.

Every subject that calls for interviews will have its own unique opportunities and challenges. Most such interviews will not be as contentious as those that cause controversies and hard feelings among otherwise friendly anglers. But all interviews should be approached with the same conscientious professionalism.

Fishing and Telling

Be very careful how you handle "destination" writing. Travel writing has always been a significant part of fly-fishing publications, and done right it's an honorable part of the tradition. Readers love such stories even if they have no intention of going to the places described; there's great vicarious fun to be had just in going along as a reader. Modern fly fishing's leading travel writers, such as the late Ernest Schwiebert, whose popular books *Remembrances of Rivers Past* (1972), *Death of a Riverkeeper* (1980) and *A River for Christmas* (1980) were vivid chronicles of his fly-fishing adventures all over the world, taking us to remote destinations that few of us can ever hope to visit in person. Their stories bring to life the amazing richness and diversity of the fly fisher's world.

Setting aside the joys of reading such stories, we encourage you to be careful. We've already talked about the risks of being hired by a lodge owner, by the magazine, or by any other party that has a vested stake in your story being positive, even if your experience wasn't. We won't belabor that challenge here. But there's another set of issues facing you: Those faraway places don't exist in a vacuum, just waiting for you to come and fill it with your own adventures. One of the least savory parts of our literary tradition is the "kiss-and-tell" fishing writer. Fifty years ago, the novelist and master fly-fishing storyteller Robert Traver called them "compulsive squealers," which is much kinder than many of the other things they've been called.

The locals along a stream—and in fact anyone who has discovered and enjoyed the fishing of that stream before your article appeared—can get pretty proprietary about their home waters, especially if there's a sudden irruption of out-of-state plates in the parking areas and a corresponding crowding in the best pools and runs. This isn't a simple matter, of course, as usually you're legally entitled to write whatever you choose about any place you choose, just as those new fishermen are legally entitled to fish those spots. Besides, there are invariably other local interests—guides, lodges, tackle shops, gas stations, and so on—who couldn't be happier about all those out-of-state plates.

The late Bud Lilly, for many years one of Montana's foremost outfitters and tackle dealers (Tom Brokaw described him as "a founding father of modern Western trout fishing"), was confronted every day with the complications of both promoting and protecting the famous trout waters of the Yellowstone country. He recognized the irony that he was not only giving advice, but selling other people's books, which directed the newcomer to just the right spot on nearly a rock-by-rock basis, while there were other anglers who'd spent years of serious fishing to find each rock and thus to earn the fish behind it. Bud recalled longtime local anglers coming into his shop, seeing a stack of kiss-and-tell books for sale, and saying, "I think I'll buy six copies and burn 'em." You don't have to approve of that sort of hostility, but you need to know it's out there.

What we're counseling here is a thoughtful, sympathetic approach to writing such stories. It's a good idea to start by asking yourself some questions.

First, and probably most important, can the fishery you're writing about stand the increase in fishing pressure? Fisheries managers call what you're doing "hot-spotting," that is, dramatically increasing the fishing pressure on a place, with all the predictable and potentially substantial social and ecological consequences. Some would say that the effects of such hot-spotting are as short-lived as your article, but others would argue that once hot-spotted, that water will never return to its pre-exposure level of use. Whether this is a bad thing or maybe even a good thing—after all, every water we fish needs as many friends as it can get—will probably depend in good part on where you live. But however ambivalent and even irresolvable such matters are, we still encourage you to proceed cautiously.

Then, and this makes it as personal to you as it is to the locals, how would *you* feel if it was your home river and suddenly all your favorite spots were full of tourists? On the one hand, they have just as much right to fish there as you do. On the other hand (and there's always another hand, sometimes several), you may have spent the past 20 years learning your stream, reach by reach, until you have every right to feel that you've earned the fish that some far-off writer is now handing to all these strangers on a platter.

Who knows, maybe if you think hard enough about all this, talk to the right people and get around to some busy streams, you can even get an article or two for your efforts.

Formatting Your Manuscript

Before you write your first story, you need to establish the page format for your manuscript. Some magazines specify formats in their submission guidelines, and if your target magazine is one of those you should follow its specified format. Otherwise, use a generic page format like the following:

Margins: Set your margins to leave at least 1¼ inch of space at the top of each page—enough room to allow room for a header. Set the horizontal margins of each page to leave approximately one inch of space on both the left- and right-hand sides of the page.

Header: Your manuscript should include a header repeated on each page. The header should include your name, contact information (email address and/or telephone number), the story title or other identification, and the number of the page currently displayed on your computer screen. All this information is necessary in case manuscript pages (either as computer files or printed pages) should accidentally become separated by the publisher. The first page also should include the total word count of the manuscript.

The header may be displayed on either the top left-hand or top right-hand corner of each page, including the first page. Some writers prefer to list page numbers at the bottom center of each page instead of in the header, and this also is satisfactory. Most "word-processing" programs include functions to determine the total word count of the manuscript and some also have programs to set up headers that automatically track page numbers and add them in consecutive order so you don't have to do it manually; however, these programs can be difficult to use and are seldom well-documented. If the story is a short one—three or four pages—it might be easier to compose the headers manually.

Indents and line spacing: Indent each paragraph of your manuscript at least five spaces (most programs make it possible to specify the number of spaces in the indent and automatically repeat it at the beginning of each paragraph). As for spacing, set your computer to add at least 1½ or two spaces between each line of type.

This generic format will serve for book manuscripts as well as magazine stories. Once you get it set up, you're finally ready to write!

Story Structure

Imagine for a moment you arc the jaded editor of a fly-fishing magazine. You spend many days sifting through unsolicited submissions looking for something you might be able to use, and you've seen just about everything. Despite your magazine's obvious devotion to fly fishing, you still get stories about fishing with worms, salmon eggs, crickets or other unmentionable things. You still see lots of what editors call "Me and Joe Went Fishin'" stories, mindless, poorly written clichés that wouldn't pass muster in a fourth-grade class.

Of course, these days some editors won't even look at unsolicited manuscripts, but we're pretending here that you're one who does, hoping to find an as-yet-undiscovered writer of great talent and promise. Still, you don't have the time or stomach to read each story from beginning to end, so instead you read only the first paragraph; often that's enough to tell you all you need to know, and the story usually ends up on the reject pile. There are times when you despair of ever seeing an opening paragraph that's strikingly original, different, or otherwise compelling, an irresistible invitation to continue reading—the kind of opening that makes an editor's day.

One also likely to make a writer's day.

This means the first paragraph of any story you write is *absolutely critical* to its success. It's the most important part of the whole story, the few words that will likely determine the story's fate.

Editors usually refer to that vital first paragraph as the "lead." It was once axiomatic that a lead paragraph should be no longer than 35 words, a sort of arbitrary assumption that the average reader (or editor) has a short attention span. That axiom has pretty much passed out of use, but it's still a good idea to keep the lead brief.

Because the lead paragraph is so vital to the success of a magazine story, we're going to spend a little more time examining types of leads and whether they succeeded or failed, starting with a short quiz. Following are the leads of three fly-fishing stories that were actually submitted for publication. Read them and decide which one(s) you think were accepted:

Lead No. 1: "I wouldn't lie to you."

Lead No. 2: "The immigrants, crowded into the damp hold of the German steamship *Werra*, were not particularly welcome when they made landfall in the United States on February 24, 1883. Xenophobic feelings were running high, with many Americans worried that the Europeans would displace residents already struggling to stay afloat. The critics were quite nasty about the newcomers, variously described as scaly, voracious, monstrous and homely. They stole food from natives. They had sharp teeth. They ate their young. They were greenish yellow with red spots. They were fish."

Lead No. 3: "The fish tore through the limbs of the submerged tree and as soon as he exited the other side, he took to the air. It was over so quickly that I just stood there staring at the fading rings that only moments before had been a silvery, leaping fish."

Answers: You may be surprised to learn that all three of these stories were accepted for publication, although, as you'll see, there was never any doubt about one of them.

The first example was the lead paragraph of a story accepted for publication by a leading fly-fishing magazine. It was written by the late Lefty Kreh, whose reputation as an angling writer was known to almost every fly fisher, and it's about as close to perfect as a lead paragraph can be. That's because (1) it's about as brief as you can get; no need to wander through excess verbiage to get to the point here, and (2) it sets the hook immediately and makes the editor or reader want to keep reading. Why? Because he or she will want to find out what it was that Lefty wasn't going to lie about.

The second example was accepted for publication by a high-profile non-angling magazine. It's a good example of what's called a misdirection lead, one that points readers in one direction and teases them a little, making them first assume the story is about human immigrants, then surprises them at the end by revealing the story is actually about fish—brown trout in this case. By the time readers get that far, they're usually hooked.

If this lead has a fault, it's that the reader has to stick with it for some distance (85 words) before the truth is revealed. Some readers (and editors) don't have that much patience, and that's one reason most lead paragraphs should be brief. Apparently, in this example, the editors weren't too worried about that, but most editors probably would have shortened the lead.

The third example was "accepted" for publication on the back side of a fishing calendar, but the person who wrote it also was the publisher of the calendar, so "acceptance" in this case was a foregone conclusion. The back side of a calendar was a good place for it, because it has so many things wrong it almost certainly would never have been accepted by any other publication. To begin with, the scene it attempts to describe is a cliché; countless fishing stories have begun with similar accounts of a big fish quickly lost. It's also sexist; why did the writer assume the fish was a male? It would have been far better to refer to the fish as a gender-neutral "it." Finally, the wording is awkward. You probably understood what the writer was trying to say, but somebody with little fishing experience might have trouble figuring out how that "silvery, leaping fish" had suddenly become "fading rings."

The lesson here: Avoid clichés, sexism, and awkwardness. Also, in writing, you *can't assume anything about the level of your readers' awareness of the subject.* It's smart to write considering the possibility that the reader might be totally unfamiliar with your subject, which means you must always make it perfectly clear and complete.

And remember: The objective of the lead paragraph is to make the reader want to read the next paragraph, and all that follow.

The Rest of the Story

Your attention-grabbing lead should be followed by explanatory paragraphs that set forth the theme or premise of the article, followed by a narrative of the remainder of the tale and a logical conclusion that supports the narrative. (There is an old sarcasm about writing how-to stories, whatever the subject: first you tell 'em what you're going to say, then you say it, then you tell 'em what you said. Try not to be that cut and dried about it, but it does suggest a kind of coarse outline for how many stories go.) Until you've written enough stories that their usual framework becomes second nature, it can be helpful to draft an outline first, then follow it as you construct the story. The outline should include key points of the story with supporting information or documentation, examples or personal experiences supporting the objective of the story, then finally transition to an appropriate and logical conclusion. Depending on the type and purpose of the story, the conclusion might be humorous, poignant or otherwise emotional.

On the way to your conclusion there are several techniques you can and should use to make your narrative seem more attractive or alive. One is to keep your prose uncomplicated, saying things in the simplest way possible so there can never be any doubt about your meaning. Avoid jargon. These seem like obvious things for a writer to do, yet it can often be difficult; even the most experienced among us sometimes struggle to make things easy for the reader, which always should be the goal of any writer. Writing in a simple, uncomplicated fashion may, in fact, be more a function of natural talent than any amount of learning or experience.

Another useful technique is the use of so-called action words. That means nothing wimpy in your choice of descriptive terms; instead, use hard-hitting words like "strong," "live," "bold" or "brilliant," perhaps linking them in alliterative fashion.

Describe scenes as they appear in your own memory; that will trigger memories in the mind's eye of the reader, adding realism and life to your narrative.

If you can customarily blend all these elements together, as though part of a musical score, you will have gone far toward establishing your own literary voice.

Just to make sure your readers are paying attention, it's also sometimes a good idea to surprise them a little. Misdirection is one way to do this, leading readers in one direction only to make a sudden turn at the end, as in the misdirection lead cited above. Another way is to point them toward an expected conclusion only to spring a different one at the end. Don't do this too often, however; just frequently enough to keep them on their reading toes.

Here's another approach to getting your story started, one that goes against most writing guidelines but may work for you, especially if you've stalled out before you even got going. Forget about structure, about beginnings and endings and transitions and all that; just tell the part of the story that you already have in your head. Usually there's some seed of idea or experience that got you thinking about the story in the first place; go ahead and write that, then work out from there. This serves at least two good purposes; first, it gets you off the starting line and under way, and second, it firmly plants in your mind what's the most important message you want to share. You may be surprised—Paul, who has occasionally used this approach, often is—to find that after spending some time elaborating forward and backward from the bit that's really the heart of the story, he suddenly realizes he's written the whole piece, or at least all the key elements, and can now fine-tune them as need be.

And don't forget: When you think you're finished, let the story simmer at least a few days and then have another look. You'll see changes you want to make, and after you make them let it simmer again for a while and take another look. Repeat as necessary.

Getting It Right

It all sounds pretty simple, right? But before you even *think* about submitting a story to a fly-fishing magazine, you need to practice, practice and practice your writing, then practice it some more. You also need to subject it to criticism from others, asking them to be as brutally frank as possible. You'll probably hear things that never occurred to you, and they may turn out to be very helpful. Listening to criticism will require a thick skin on your part, so you must learn not to take it personally.

Where can you get such criticism? You can ask your spouse or significant other to read your copy and offer comments, though that's not always a recipe for good relationships. Same for your fishing partners. But the best way is probably to find and join a local writers' group. Nearly every community has at least one such group and larger communities often have several. A quick internet search will probably turn up several in your own community.

The members of these groups typically include people just like you, trying to learn the art and discipline of writing. They exchange manuscripts (or read them in group meetings) expressly for the purpose of criticism, which means you get to be a critic yourself, as well as a target. But don't ever be snarky; remember, the purpose of the exercise is to help one another.

Final thoughts: Use your spelling checker. Then use it again. Then use it one more time. But even then, don't count on it alone; read and reread the manuscript looking for spell-check glitches.

Do whatever research is necessary to make certain. Remember, however: Whatever you write, and whomever you write it for, always avoid sexism.

Also remember to remind your readers occasionally that fly fishing isn't "free;" as a participant, you have an obligation to "give back" to the sport, which means following good conservation practices and contributing to organizations trying to preserve or enhance fisheries habit.

And don't forget to observe proper angling etiquette, in your writing as well as your fishing.

Avoid plagiarism at all costs.

When you're writing about fly fishing it's also very important to describe the people and places you're writing about—histories, conversations, physical surroundings, etc.—to give the reader a complete sense of your experience. It's also legitimate and often effective to describe your emotions, such as the awe you sometimes feel from the experience of fishing, or the impact of your physical surroundings. Readers can identify with such feelings.

Keep your story focused. That hardly seems like necessary advice, but you'd be surprised how many writers have trouble keeping their eye on the fly. Steve once had to edit a story that appeared at least three times longer than the space available in the magazine for which it was intended, and it looked as if it would be a tough job to cut it down to size. It turned out to be just the opposite, however, because the author—an experienced writer who should have known better—completely lost track of the story he was supposed to be telling. He had spent a week at a fishing resort owned by a woman, but this was no ordinary woman; she had long, blond hair, a perfect face, inviting lips, a regal figure, a laugh like tinkling bells, and she was a brilliant conversationalist who spoke several languages. She also was a gourmet chef who turned out incredible meals night after night, singing sweetly as she worked. The resort staff worshipped and respected her as the perfect boss. And so on and on.

It was obvious the author had become completely smitten with this incredible woman, to the point he hardly even mentioned fishing in his story until the very end, where he added a few paragraphs almost as an afterthought. It was easy to delete everything in the story except those brief concluding paragraphs, leaving only a single mention of the woman's name as owner of the resort, and once those trims were made the story fit the available magazine space exactly. It seemed a little heartless to leave this budding affair on the cutting-room floor, as it were, but it had to be done because the story was for a fly-fishing magazine, not *True Romance.*

Another writer, this one less experienced, similarly lost track of the story he was trying to tell. He was fishing in an area with lots of old military fortifications, which fascinated him to the point that he began describing them at great length and exhaustive detail, making the story impossibly long. However, as in the previous example, he evidently finally remembered he was supposed to be writing about the fishing and added just enough information to make the story publishable after his descriptions of all the forts, bunkers and earthworks were removed. Again, the story was for a fly-fishing magazine, not *Soldier of Fortune.*

Get your facts straight. As editor of *The Flyfisher* magazine, Steve established a policy of trying to publish a fictional story in each issue, a goal that proved tough to reach consistently. Among the unsolicited manuscripts he received was one that looked promising; it was well written and had a plot that seemed to work, a combination seldom found in unsolicited works of fiction. But it had one whopping error: A vital part of the story concerned spawning brook trout, and the author had them spawning in the spring.

Brook trout spawn in the fall.

Ordinarily, that mistake alone would have been enough to consign the manuscript to the reject pile. But decent fly-fishing fiction is hard to come by, so it seemed worth the effort to try to save the story. This was done by reversing the seasons and making other changes to reconcile the seasonal change, and the story was finally published. Many editors probably wouldn't have gone to the trouble to resurrect such a flawed story, which just goes to show that a single mistake can make all the difference in acceptance or rejection of a story, whether fiction or nonfiction.

Don't let that happen to you.

Writer's Homework Assignment

It hardly seems fair to let you off the hook merely by having you read all the advice we've just given in this chapter, so we're going to suggest some good practice to sharpen your skills. This isn't a casual exercise. It can improve your ability to handle a story significantly. Here's what to do:

1. Write a story approximately 2,000 words long. Read it out loud to yourself to see how it sounds; you'll undoubtedly hear things you want to change. Let it sit for a while, then have another look; you'll see more things you want to change. Keep polishing and refining it until you're absolutely satisfied it's your very best work.
2. Now cut the story to 1,500 words. Don't fudge! Less than 1,500 words is OK, but not one word more than 1,500!
3. When you've finished doing that, cut the story to 1,000 words. Same rules apply.
4. After you've finished cutting the story to 1,000 words, cut it to 750 words. Same rules again.

Repeat all this with different stories as often as possible, then repeat it some more. After the first story, which you must write yourself, you can save time by using articles from magazines or the internet (just make sure you destroy all versions when you're finished). But for best results, use your own work as often as you can. If you keep doing this over a long time—we're talking several years at least—it will teach you how to organize material, how to decide what's important and what isn't, and how to use words sparingly—and those are essentials of good writing. Except for professional training and experience, it's the best way to learn how to become an accomplished writer.

Why go through all this? Well, you wouldn't expect to take your first-ever turn at bat in the World Series, would you? You have to play in the minors and work your way up if you want to play in the big leagues. So . . . batter up!

An authority who lays down a law and dogmatizes is a narcotic, a soporific, an opiate. The true function of an authority is to stimulate, not to paralyze, original thinking.
—G. E. M. Skues, *The Way of a Trout with a Fly* (1921)

CHAPTER SIX

Telling Your Story

Maybe you already have a story you're anxious to tell; that's a good place to start. But it's not unusual for writers to have a challenging time coming up with ideas for articles. If you need a little help on that score, this chapter offers suggestions about where to find ideas and how to develop them—plus a few ideas you should probably avoid.

Some of the best ideas are likely to come from your own experiences. As a fly fisher, you've probably had some memorable trips and adventures; think about those and one or two likely will emerge as good candidates for a magazine story. For example, Steve vividly recalls getting caught in a volcanic eruption on a fishing trip—something that surely doesn't happen every day!—and the experience resulted in a lucrative magazine story, followed by an invitation from the magazine to return to the scene of the eruption a year later and write a follow-up tale that turned out to be a cover story.

Another of Steve's trips ended in a middle-of-the-night evacuation from a remote fishing camp in the path of a rapidly approaching wildfire. An even less pleasant occurrence was an excruciatingly painful kidney stone attack on a hard-to-reach water that ended in a long ambulance trip to the nearest hospital. Remember the old saying about clouds and silver linings, though; such experiences can make good material for magazine stories (or book chapters).

Hearing of these interrupted trips of Steve's reminds Paul that during his long residence in Yellowstone Park he had to change his plans for the day—or even call it day before he meant to—because of encounters with the park's other residents, especially the ones a lot larger than Paul. More than once he made quick retreats from hostile elk, or made wide detours around bison and moose, or just turned around and went home because a grizzly bear had moved into a favorite reach of stream. Realizing this now, Paul thanks Steve for reminding him of some entertaining if not harrowing stories that have so far remained unwritten.

Sometimes, though, the unusual or striking element of a story may be on a much more modest scale, such as the time Paul, fishing late into the evening on a favorite Wyoming stream, hooked a bat on his backcast. He hadn't been paying enough attention to the abundance of bats feeding on the same hatch as the trout, until one bat zeroed in on his fly—which says a lot about how fast bats are when chasing prey. For a long time this was just a creepy memory, but eventually Paul's little experience worked its way into an article called "The Collateral Catch."

As these episodes prove, there are almost endless ways to come at the fly fisher's world. Fly-fishing history is always a good and endlessly rich subject, as Paul can testify. There are also many different ways to approach it. One way is to interview or profile a famous fly fisher (or more than one), fly tyer, rod builder, author, etc. For some good examples of the form, see Jack W. Berryman's 2006 book *Fly Fishing Pioneers and Legends of the Northwest*, George Reiger's massive *Profiles in Saltwater Angling* (1973), and Martin Keane's *Classic Rods and Rodmakers* 1996). A good many such books, including Berryman's and Reiger's, began as a series of magazine articles, which reminds us to point out that it never hurts, when writing an article, to wonder how it might eventually fit into a book. Never dump computer files of your articles; you never can tell when they'll come in handy and be revived for a second life as a chapter in a book (see chapter 8).

Another idea is to write a tribute to a fly fisher who influenced your life, such as Jim McLennan's warm, well-written account of his mentor, the late Leigh Perkins, who built the Orvis Company into a major manufacturer and vendor of fly-fishing tackle and outdoor gear and engineered the contribution of many millions of dollars to conservation programs. Look at it for a model of the form: it appeared in the October–December 2021 issue of *Fly Fisherman* magazine.

Yet another approach is to profile a famous fishing venue—hotel, camp, Irish castle, etc. A fine example is Ernest Schwiebert's "Homage to Henryville," which appeared in *American Trout Fishing*, the 1966 trade edition of a book originally published in a limited edition as *The Gordon Garland*, by "Theodore Gordon and a Company of Anglers." Later, Schwiebert included the same story in his own book, *Remembrances of Rivers Past* (1972), and eventually expanded it into a full-length limited-edition book, *The Henryville Flyfishers: A Chronicle of American Fly Fishing* (1998).

It's been done before, but another idea is to fashion your own unique version of how to assemble a "complete" fishing library, and what books you think should be in it. Arnold Gingrich was among the first modern fly-fishing writers to do this, in his engaging memoir, *The Well-Tempered Angler* (1965), but several other writers have done it since. You might be making a fairly original contribution to this sort of writing if you narrow the list down to some specific topic, such as, say, ten essential books for New England trout fishing, or how to assemble a one-shelf library of bonefishing books.

Angling etiquette is also a potential subject for an article because etiquette varies from stream to stream and is evolving. Just making readers understand that things like manners and ethics are not static, set-in-stone matters is always a good enterprise for writers.

Writing about places is an honorable and, we think, inexhaustible topic. One of the biggest changes in the sport in the past half-century has been a dramatic increase in the number of anglers traveling to far-off destinations to fish waters that were until recently little known or not known at all. That exploration has been paralleled by another, identifying species of fish that were either previously uncaught or only rarely so. If you can bear the formidable expenses involved in getting to such distant places and such unfamiliar fish, don't hesitate to have a go at them.

Steve recently became acquainted with the first person he's ever met who went fly fishing in Greenland, where he found spectacular fishing for sea-run arctic char and brought back great photos to prove it. But when Steve suggested his experience would make a great story for a magazine, the angler shrugged it off on grounds he wasn't a writer. So there's a story idea just waiting for you; all you have to do is figure out how to get to Greenland when the arctic char are running.

Casting and Fly Tying

Articles about fly casting and tying methods are always in demand by magazines, so long as you don't repeat something the magazine has already published. But if you're writing about fly casting, you already might have discovered you've taken on the toughest of all fly-fishing topics. That's because it's just plain downright difficult to find the right words to describe the simultaneous human movements required to cast a fly, or what those movements should feel or look like. There's no easy way to do it, and lots of writers have found that out the hard way.

Just think about it: If you're trying to describe fly casting from a right-hander's point of view, then what is the left arm doing while the right arm is busy with the casting stroke? How are those movements coordinated? What should the timing be? How much tension should the left hand keep on the line? What are the feet doing in the meantime? How should their movements be coordinated with everything else? Or should they be? While all this is happening, where are the eyes looking? Fore or aft? And perhaps most important of all, what is the brain thinking? Finding the words to relate and coordinate all these things is a hugely difficult task, made even more complicated by the idiosyncrasies of the individual trying to do it. It doesn't make it any easier that when you've figured out the best descriptions for right-handers, you also need to translate them for left-handers.

Good luck.

So before you even think about writing about fly casting, you should be a highly competent and experienced caster yourself. That doesn't mean you need to be a tournament champion, but it does mean you should be accomplished at handling rods of any length and action and lines of any weight and taper. The double haul should be as familiar as looking at your wristwatch, and the roll cast and other specialized casts should be established parts of your casting repertoire. And these days you need to be equally conversant with Spey casting.

Keep in mind, however, that if you're trying to write something that will help a true beginner learn to cast, you need to stick to describing only what it will take to get a fly 40 or 50 feet in front of the caster in a reasonably straight line; the double haul, roll cast and other specialized casts can wait until the learner has some experience.

Is there anything that can be done to make it easier to describe the basics of casting? Well, there are no guarantees, but one possibly helpful suggestion is to start by trying to explain (in writing) the physics of fly casting—how the rod works as a lever to propel the fly line and the fly to the intended target. If you can convey this clearly so people understand how the process actually works, it may help them grasp the remainder of the concept.

This also is a case where you definitely should seek the criticism of others, preferably people who have never been fly fishing. You might even have to provide them with a rigged-up fly rod, reel and line to see if they can figure out what to do as they read or listen to your instructions. Starting with such an "audience of innocents" should yield some immediate feedback indicating whether your written account is getting the idea across.

Where to get such feedback? Your local writing group, previously mentioned, could be one source. Another might be a local Scout troop; the Scouts now have a fly-fishing merit badge, so they should be learning how to cast in any case. Failing that, maybe you could convince the physical education teacher at your local high school to devote part of a day's class to casting instruction.

If you pay close attention and take notes from any or all these sources, you may end up with some very helpful insights that will greatly improve your descriptions and presentation. Then, and only then, should you solicit feedback from an audience of people with at least minimal experience with a fly rod, and your local fly-fishing club is probably the best place to find them.

If you're writing about fly tying, then take a big sigh of relief; it's a lot easier than trying to describe casting. That's not to say it isn't still complicated; you need to describe clearly each step of the tying process for each fly, explain the basic tying motions, relate the properties of each material involved in the pattern, the proportions needed, and how to manipulate those materials. You must tell how to maintain tension on the tying thread, what knots to tie, what tools to use and how to use them, and describe a seemingly limitless number of other little or big steps necessary to finish a fly that will be as pleasing to the fish as it is pleasing to the eye. That's a lot of information to get across for even a simple, basic fly pattern—let alone a classic Atlantic salmon fly.

Be thankful there are scads of fly-tying books available to help you or serve as examples. Here are just a few: *The Fly Tyer's Primer*, *Mastering the Art of Fly Tying*, and *Modern Fly-Tying Materials*, all by Richard W. Talleur; *Flytying Techniques* by Jacqueline Wakeford, and *Fly-Dressing Materials* by John Veniard.

If your story is about a new fly pattern, you need to be certain it *is* new, not one that has already been published in magazines or books. That could be more difficult than you think, because few ideas in fly fishing are really new, and truly original fly patterns are relatively uncommon. So do some research before you offer your discovery to a fly-fishing magazine; you may find someone else has beaten you to the punch.

Keeping Track

As we mentioned briefly in chapter 2, there are few things you can do to stimulate writing ideas more important than keeping a journal or diary. Whether it's just a few scribbled notes about what seemed most important in that day's fishing, or a disciplined log that records the date and location, the weather, hatches or other fish prey seen, timing of feeding activity, fly patterns used (including which worked and which didn't); size, species, and numbers of fish caught; birds sighted, and companions and notes on how they did. If you're not doing it already we urge you to start—not now but *right now*.

Imagine living the lovely cliche of sitting down at the end of a day's fishing—perhaps in the traditional stereotype of an angler, with a glass of something potent at hand, lounging in front of a picturesque fireplace, with a favorite retriever snoring gently nearby, and reliving the day by writing up your notes and impressions. We know writers who put down long, polished narratives in their daily journals. Others just scribble a few notes on the back of a handy grocery list or any other scrap paper within reach. Whatever you decide is right for you, make sure it's never so time-consuming that you resist doing it.

There are good reasons for keeping a journal that go way beyond the joys of remembering by the fireplace. Do it for reasons you can't even imagine yet. You're storing information whose value might not become evident for years, and you can't yet tell specifically what it will turn out to be. Didn't that hatch used to happen earlier than this? At what level of the river has this stretch been most productive? Is it just my imagination, or do the trout not begin feeding until the sun is off/on the water? What did that guy in the goofy hat say about streamers that day on the Upper West Branch? Wasn't it just two years ago that I bought these waders that are already leaking like the White House press corps? Such information will not only help you fish more intelligently in subsequent years, but you might be surprised how often it may inspire the idea for an article. All you need to do is reread your journals every now and then.

Paul's big brother (also named Steve), who started tying flies about 70 years ago, cooked up an all-purpose Clouser-style saltwater fly pattern he jokingly called the Pink Wonder. As his saltwater experience piled up, he became almost exclusively devoted to that fly, using it in a hundred places and always catching his fair share of fish. For his own entertainment, he kept track all that time, and eventually had a list of 77 species of salt- and freshwater fish he'd taken with that one fly pattern. Eventually he wrote a little article about the fly for a magazine, but the point of the episode is that such long, patient experiences can provide opportunities for articles if we are of a mind to write them. Imagine: seventy-seven species! How many places must you go to find that diversity of fly-fishing quarry? Which ones were the largest/smallest/most difficult to hook/most difficult to land, or ugliest or prettiest? Well, Paul's brother could answer a lot of questions like those because he kept track. So should you.

Here's another thing about keeping track: Now that practically everybody is carrying a phone with a camera in it, don't neglect its possibilities beyond the obvious grip-and-grin trophy shots. Since his first years of fly fishing Paul has almost always had a small camera with him. It became an essential tool for him far beyond the usual fishing scenes, enabling him to take thousands (no kidding) of what he thought of as "visual field notes," artless photos of something he just wanted to remember—a rarely seen wildflower, some beaver chewings along a stream, a bear track, a moose he wasn't yet sure was going to let him go by on the trail, and all the other little things that bring a day's best moments to life in memory. He kept track, and nowadays, what with almost everybody carrying a phone camera, there's no excuse for not easily making a visual record of the day (many of these phones have incredible macros, so don't miss a chance to take home a sharp, even publishable, close-up image of whatever naturals are hatching).

A well-kept fishing diary also can sometimes help correct a faulty memory, as Steve can testify. After writing in two different books that he landed five large trout in his first pass through the famous Major Jones Pool of New Zealand's Tongariro River, he happened to look back at his diary entry for that day and discovered to his chagrin that he'd really landed only three; the two others were hooked and lost. Which means it's always a good idea to check your diary before you write, not afterward.

And one last tip: Keep a tablet and pen next to your bed in case you wake up with a good idea in the middle of the night; jot it down and see if it stands the test of daylight in the morning.

Stories to Avoid

As long as we're on the subject of story ideas, it's also worth mentioning a few you should probably try to avoid. Many fly fishers have deep philosophical or moral objections to fishing contests or tournaments, believing firmly that fly fishing and competition do not mix, so those topics are best avoided. The same is true for stories about fishing records or anglers who call themselves "professional fly fishers;" they are better left alone. And regardless of your topic, you should always avoid being argumentative, hostile or arrogant.

Similarly, you should strive to avoid writing the same stories that others have written many times before you thought of them. For example, jaded readers (like us) of fishing stories will urge you never to start an article or story with a lurid account of hooking a great fish on a fly and fighting it under almost impossible circumstances only to lose it at the very last moment and end up feeling as if your best friend has just died; it's been done before, so many times it's become a cliché. There are other subjects and stories that have been done to death, and you should avoid those, too.

Different eras seem to have had their own worn-out storylines, but clichéd stories have been a laughing matter among American fishing writers at least since 1878. That year, in his vastly entertaining if now forgotten book *In the Wilderness*, essayist and novelist Charles Dudley Warner published *A Fight with a Trout*, his delicious parody of the overblown tale of catching a ferocious fish that was even then a chestnut in outdoor writing. (It was apparently Warner, by the way, and not his friend Mark Twain, who originally said, "Everybody complains about the weather, but nobody does anything about it.") We cannot resist a brief quote that captures the fun of his parody, which involved an epic struggle with a trout that, when finally landed, weighed three-quarters of a pound: "The trout left the water about ten feet from the boat, and came directly at me with fiery eyes, his speckled sides flashing like a meteor. I dodged as he whisked by with a vicious slap of his bifurcated tail, and nearly upset the boat."

Warner concluded that "Fish always lose by being 'got in and dressed.' It is best to weigh them while they are in the water. The only really large one I ever caught got away with my leader when I first struck him. He weighed ten pounds."

For another example from the crop of literary chestnuts, consider tales of the fancy-pants angler outfitted with all the best equipment who is outfished by the barefoot boy with a pin on a string; these tales often concluded with the man buying the fish and taking it home as if he caught it.

(There are times, however, when life can imitate a well-worn theme. Steve remembers a fishing-camp owner who rented a cabin to a flashy, big-city attorney making his first fly-fishing trip. The attorney arrived dressed to the hilt in the latest fishing fashions and weighted down with expensive tackle and accessories; the only thing he lacked was outdoor experience. The woodstove in his cabin baffled him; he'd never seen one before and ended up building a fire in the stove's oven instead its firebox, then nearly suffocated in the smoke. Next day he went home.)

Another of the most tiresome stories—at least before we became a little wiser about such things as gender stereotypes—told of someone's girlfriend (even worse, "little woman") invading the bachelors-only fishing camp, putting up curtains, disrupting the drinking and unhealthy eating, and then casually or accidentally catching the biggest fish.

There are also abundant stories about a giant old trout (usually named "Grumpy" or "Mossback," or something like that) that has avoided the efforts to catch it that you and your fishing companions have been trying for years—no matter who catches it in the end—just as there's a pretty good backlog of stories about someone meeting an ancient angler who's obviously on his last fishing trip and leaves you with his favorite Hardy reel, bamboo rod, or a box of impeccably tied flies. Another worn-out tale, usually characterized as the "Me and Joe Went Fishin" story, has been thrashed over so many times as to become almost unreadable for many experienced readers and for most editors.

Nevertheless, sometimes it doesn't hurt to keep some of these old chestnuts in mind; there still may be life in them. Suppose the lurid account involves hooking a species of fish that we jaded readers and editors have never even heard of? Or if it turns out that the fisherman being described is a six-year-old who built the rod and tied the fly, or is just a regular angler who's had to endure great hardship to reach some remote water, or to overcome some severe disability just to hold a fly rod? What if the fisherman is the first to fish the stream, which has been closed to angling for many years or was drowned for a century under a reservoir whose dam has finally been removed, allowing the return of long-absent runs of migratory trout or salmon? Such circumstances can bring a brand-new perspective to a well-worn subject, and an alert writer should always be watching for these "what-ifs."

Got all that?

Writing for Real

Now it's time to start writing for real, applying all the practice, preparation and learning we've been through up to this point. By now we hope you're bursting with story ideas and ready to commit the best of them to print. You should have a target magazine in mind, one you've studied thoroughly so you know what material it needs and what it has published recently.

So it's time to go to work.

Before you type that first word, however, remember to set up the generic format we discussed back in chapter 5 on page 90. Remember, though, some magazines specify their own formats; if you're writing for one of them, use theirs.

After you've got the format established, next comes what we've emphasized is the most important paragraph of all: the lead. Maybe you've tossed that around in your head and already decided exactly what you're going to say—or maybe not. Either way, that vital first paragraph will probably need at least a few adjustments until it's the brief, sparkling, original, irresistible summons to editors and readers that it needs to be.

Once you're past that important hurdle you can settle down and begin fleshing out the rest of the story, refreshing your mind, if necessary, by consulting the "Story Structure" section in chapter 5.

May the words be with you.

Those authors who would find many readers, must endeavour to please while they instruct.

—Samuel Johnson

CHAPTER SEVEN

Selling Your Story

You should be feeling pretty good right now because you've just finished writing your first story for a fly-fishing magazine. All that hard work is finally over, right?

Well, not quite. Next comes the *really* hard part: Finding a magazine that will buy and publish your story. This chapter will tell you how to go about it.

Most writers want to see their first story published in their favorite fly-fishing magazine, and that's natural, but as we've emphasized repeatedly in these pages your first priority is to make sure your story is a good fit for that publication. If your story is about a new size 18 no-hackle mayfly pattern, a magazine devoted to Spey casting isn't going to buy it. So do some research and make certain you choose a target where your story *will* fit comfortably; it may not be your favorite magazine, but it absolutely should be one likely to accept the article you've written.

You should also consider how your chosen magazine pays its writers. Some pay on publication, others on acceptance, and it's always best to work with one of the latter. The most obvious reason is that you'll get paid right away; if you have to wait until your article is published, it might be a long while before you see any money. Another reason is that payment on acceptance means a magazine has made at least a small investment in your article, which gives you a little leverage to push back if an editor decides something in the story needs to be changed and you'd rather not change it. As a rule of thumb, however, you should always try to respond positively if an editor asks you to change something; he or she almost certainly has a good reason for doing so, even if it's not immediately obvious to you.

Sometimes other things can go wrong if you're dealing with a pay-on-publication magazine. One of Steve's first sales was to such a magazine, which published the story, but the magazine's check wasn't in the mail. After waiting a reasonable period, Steve called the magazine only to get a recorded message that its phone had been disconnected. Inquiries led to the news that the magazine had gone out of business immediately after publishing the issue that contained his article, defaulting on the $50 fee it promised. Back then that seemed like a lot of money.

Steve's case was unusual, though it's still not completely unheard of for a magazine to disappear suddenly without paying its writers. More often, a magazine may accept a story and later decide not to publish it, usually for reasons beyond the writer's control. When that happens, some magazines offer what's known as a "kill fee," usually a percentage of the fee agreed upon when the story was accepted. However, kill fees are a little uncommon among fly-fishing magazines because many lack the financial resources to offer them. If your target magazine is one that does, it might say so in its submission requirements; if it doesn't, and if it's your first magazine sale, it's probably not a good idea to ask about kill fees; getting that first acceptance is the most important thing, and once your story is accepted the odds are heavily in your favor that it will be published.

So make sure you choose your target magazine carefully, not only to fit its usual content but also for its payment policies.

It also never hurts to have a backup target in mind.

To Query or Not to Query

Once you've settled on a target magazine, the next thing to consider is whether to send it a query. But just what is a query, anyway? And why is it important?

A query is a brief (usually one page) email or letter describing your proposed story in such an original or clever way that the magazine's editor will be hooked instantly, and he or she will quickly write back and ask to see the finished manuscript. The submission requirements of most magazines (if they bother to post them) often specify whether the magazine prefers to see queries instead of finished manuscripts, although some will accept either. If your target magazine prefers queries, then that's what you should submit, even if the publication also accepts finished manuscripts. Why? Because if you haven't yet finished writing the story you might save yourself a lot of time; it's much faster and easier to write a single-page letter than a full-length story. Another good reason is that a query can open online conversation if the editor likes your story idea but wants it shaped a certain way.

Query letters should always be written with patience and care. The fate of your story idea depends on the professionalism, cleverness and originality of your query, just as the fate of your story depends on its lead paragraph, so queries require plenty of thought and preparation. Yours should be designed to make an editor sit right up and exclaim, "Now that sounds like a winner! Hey Joe! Come look at this!"

Here's how to format your query and what (and what not) to put in it:

- Select an easy-to-read, 12-point typeface, such as Times New Roman or Arial. If your query is on a printed page, use high-quality white paper.
- Set one-inch margins and single-spaced lines.
- Include a header with the date and your contact information.
- Address the letter to the magazine's editor by name (you should be able to find it in the submission requirements or in the magazine itself). If you use "Mr.," "Ms." or "Mrs.," make sure you get the gender right. If you've dealt with the editor previously and feel comfortable addressing him/her by first name, do so.
- Capture the editor's attention with an irresistible opening line. Make him/her want to read the rest of the query. Don't get too carried away, though; never say something like "This is the greatest story your magazine will ever publish." That's a sure ticket to the round file.
- Follow your irresistible opening line with a brief synopsis of the proposed story. It may help if you also identify the demographics of the readers you (and the magazine) are trying to reach.
- If you're already a published writer, include a brief summary of your professional experience (emphasis on brief). Don't exaggerate.
- Do not mention what you expect (or hope) to be paid for your story.

- If you're sending your query to several magazines simultaneously, you should be up front about it; some magazines will not accept simultaneous queries. Some writers also suppose simultaneous queries make editors respond quickly as if they were competing for the story; editors, however, may take a different view. So be careful.
- Thank the editor for his/her consideration of your story. Then close the letter "sincerely" or with "best regards" (not "tight lines") and your name. Sign your name if the query is on paper.
- Proofread your query carefully. Use your spelling checker. Then use it again. Read the letter aloud to see if it reads smoothly, just as you would a manuscript. Let it sit for a while and take another look before you send it.
- Never start your query with a rhetorical question, such as "Have you ever seen a fly tied with feathers from an extinct giant moa?" Nor should you ever say anything disparaging about the target magazine, such as "I thought you really missed the target with your story about the world's best fly fishers, but my story will get you back on track." Or something like "Your magazine may not be the biggest or best, but my story will help it get there." Or "I've sent the same query to several other magazines, so if you like my proposal you'd better respond quickly." These seem like obvious examples of what not to do, but you'd be surprised at some of the crazy things that turn up in editors' inboxes.

Resist the temptation to tell the whole story, or to start your query with an anecdote. Just give the editor the basics in terms that will make him or her want a look at the story: the subject has been neglected; the place you want to write about is little known (but can stand the pressure); the fly you've devised was inspired by earlier patterns with a lively tradition that can be elaborated in the article, and so forth.

A little humility sometimes can go a long way, too. For example, don't go on the attack by saying previous writers on your subject have been idiots who got it all wrong. That's not good salesmanship, and the editor may very well have strong loyalties to some of the very writers you're claiming are idiots. Don't set yourself up as if you were doing the editor a favor by offering your story (this actually happens, never works, and may set you back permanently with that magazine).

What follows is an example of a well-written query. But lest you get excited when you read it, nobody is really writing about left-handed fly rods (at least, not yet). We just thought it would be fun to use them as an example.

Wilson Wordsmith
3231 Angler Avenue
Speck, MT 5923
(411) 027-2992
words@realfastmail.com

April 1, 2023

Mark Mywords, editor
Mainstream Fly Fishing
Mail Stop 54039
Metropolis, MI

Dear Mark:

Gary Graybeard, master rodmaker at Hot Rods, Inc., is building a radical new line of bamboo fly rods designed exclusively for left-handed casters. Lefties, he says, generate different torque than their right-handed counterparts, causing compression or release of the rod's bamboo power fibers that can hamper casting accuracy and reduce a rod's life span. He uses different tapers to eliminate these problems in his left-handed rods and has the calculations, measurements and personal testimonials to back up this novel approach.

After hours of interviews and demonstrations with Gary, I've summarized his theory in an article of 1,250 words, including details on how your rod-building readers can apply his unique approach. I'd like to submit the article for your consideration, along with a wide selection of digital color photos and several detailed diagrams. This will be my seventh story for *Mainstream*; you'll recall the last one was about using drones in fly fishing, which appeared in your November issue.

If you're interested, as I hope you will be, please let me know and I'll send the article and attachments forthwith. Thanks for reading!

Sincerely,
Wilson Wordsmith

Here's the same made-up subject in an example of how not to write a query:

Izaak Walton CXIV
P.O Box 13
Right Wing, Idaho
(511) 008-8132
izaak@realquickmail.com

April 1, 2023

Editor, *Mainstream Fly Fishing*
Mail Stop 54039
Metropolis, MI

Dear Editor:

Who says there's nothing new in fly Fishing? Gary Graybeard, reknownd proprietor-owner of Hot Rods, Inc., has come up with a totally new idea for building bamboo fly rods: He believes rods for left-handed casters should be thicker in some places and thinner in others than rods for people who cast right-handed because southpaughs have different torks. That means Gary's rods cast smoother for lefties.

I learned all this at Gary's knee, so to speak, and now I have prepared a concise article of 6,980 words that explains everything, including endorsements by several of Gary's many satisfied customers. This should be of grate interest to your readers that build or use bamboo rods. Along with the story I have lots of good photos and drawings, and I would be pleased to have you publish it for a fee of $1,500. Better hurry, though, because I am sending this same inquiry to three other magazines, and I expect they will be interested to. It would be a shame if your magazine didn't be the first to publicize this remarkable new approach.

As editor of the monthly newsletter of the Leaky Waders Fly Fishing Club in Right Wing, Idaho, I am a experienced writer. My new book, *Flies and Lies*, has also been submitted for publication. I look forward to seeing your positive reply.

Be Well,
Izaak Walton CXIV

You can find more good and bad query samples and other useful information in *Writer's Market*. Many websites also offer good advice and examples on this topic.

Rejoicing, or Rejected?

After you've written and submitted your query you need to settle down and wait for a response. This will nearly always take longer than you expected. You'd think that in this email-dominated world you should get a speedy reply, and if you're lucky, maybe you will. On the other hand, you may end up wondering if the response is coming via Pony Express—or if it's coming at all.

That was often the case back in the "old days" when it seemed to take forever for magazines or book publishers to get around to answering their mail, and sometimes it still seems that way. It wasn't that long ago Steve was surprised to receive acceptance of a book proposal he'd submitted to a publisher so many months before that he'd forgotten about it; it made him wonder if the publisher's acquisitions editor was in danger of missing her quota for signing up new book projects, or maybe had been too busy having lunch with literary agents to bother answering queries. You could write a book in the time it took her to respond.

In those days magazines were almost as bad, and their response, when or if it finally came, was usually a brief rejection slip. Rejection slips were great time-savers for editors, freeing them from the necessity of having to write individual replies. They all followed a similar formula, sometimes even using the same words; first they thanked the hopeful author for his or her "interest" in their magazine, then expressed sincere regret that "your query (or manuscript) does not fit our current needs." If ever there was a phrase calculated to cover a multitude of sins, that was it. Finally there was a brief, feel-good statement to the effect that rejection "does not necessarily imply lack of merit." Steve remembers those phrases well because as a young writer he collected plenty of rejection slips and later, as an editor, issued plenty of his own.

Today rejections sometimes may come as a quick-and-dirty email with little regard for hurt feelings at the receiving end (another reason you need a thick skin in this business). They also don't usually provide a clue as to why your query or manuscript was rejected. And once you receive a rejection, there's no appeal.

Well, hardly ever, anyway. But Steve remembers when he was editing *The Flyfisher* magazine he encountered one writer who wouldn't take no for an answer. The writer had submitted an unsolicited but well-written manuscript describing a method for splicing fly line to backing, with good photos to go with it. The only problem was that the method had already been publicized widely in magazines and books, so the manuscript was returned to the author with a rejection slip attached. Two weeks later the mail brought a small package from the same writer. Steve examined it closely to make sure no ticking sounds were coming from inside. When he couldn't hear anything, and it seemed the package didn't weigh enough to contain explosives, he finally opened it; inside he was surprised to find a pair of woman's earrings and a matching bracelet, together with a letter inviting Steve to give the jewelry to his wife or girlfriend if, in return, he would only just tell the writer the "real reason" his splicing story was rejected, "because I thought it was really good."

Steve repackaged the jewelry and sent it back to the disappointed writer with a letter saying there was no charge for advice and explained the "real reason" his story was rejected was because the method it described was old, well-known and widely used. Steve thought that would be the end of it, but a week later he received another letter from the frustrated author, who expressed effusive thanks for the explanation of why his story had been rejected. Then he added: "I just sent a story to *Popular Mechanics* that I think is really good, and if they reject it, I'll send it to you."

Gee, thanks, fella.

That was the last he ever heard from the disappointed writer. Maybe *Popular Mechanics* accepted his story.

While most magazines relied on rejection slips in those days, at least one didn't. With its weekly publishing schedule, *Sports Illustrated* always needed material, and back then it still considered fishing a sport, so Steve wrote for it regularly. But in a few instances when a story didn't meet *SI*'s "current needs," the bad news was usually conveyed in a friendly personal note from an editor, explaining the reason for the rejection and taking some of the sting out of it. Other times it was a friendly note indicating some fault with the proposed story and suggesting changes, which were duly made so the story finally made its way into print. Unfortunately, very few magazines had the staff, the time, the need for material, or the inclination to deal with writers that way, and that's still true.

Today, given the convenience and impersonality of email, about the most you can hope to receive is a curt message that says, for reasons not specified, your story didn't make the grade. And you might still have to wait a long time to receive it.

DEALING WITH REJECTION

If you've ever asked for a date and been turned down, you know how it feels to be rejected, and it hurts just as much to be rejected by a magazine. No matter how it's framed, it's never easy to accept, and it's especially tough and discouraging for a beginning writer. All too often, that first rejection is enough to cause a hopeful writer to quit the whole business in frustration and never try to write a second story. Don't be that person! Nearly every beginning writer has to suffer the pain of rejection at least once or twice before he or she scores an acceptance, which may prove to be the first of many. So don't let it get you down.

While neither of us is experienced in psychology or counseling, we can offer some advice based on experience that might help temporarily to deal with the disappointment of rejection. Sometimes it can bring immediate relief if you shout great strings of profanity, cursing the utter stupidity of the editor who failed to recognize the excellence of your story. But as soon as you've gotten that off your chest and your blood pressure is back to normal, a more useful reaction is to think of the possible reasons your story was rejected. It might have been something as simple as someone else beating you to the punch with a similar story; an internet search could provide the answer. Or, more significant, perhaps something was wrong with the story itself. Was it out of date? Again, the internet might reveal an answer. Or could it have been something you did wrong? Mistakes in spelling or grammar? Run the story through your spelling checker again, or a grammar checker if you have one. Was the story too long? Remember the writer's homework exercise in trimming manuscripts (chapter 5); perhaps you should apply that to your story; shorter is often better. Was the story poorly organized or confusing? Show it to someone whose judgment you trust and pay attention to the answer. The problem might be an obvious thing you overlooked or it could be a series of things, but it's important to track it down so you can eliminate it in your next effort—which might be revising the rejected story to eliminate the problem(s).

Sometimes, though, you just can't find anything wrong. One of the most intriguing revelations of Paul's experience as a magazine editor was that those brief little notes saying, "this doesn't meet our needs," or, "this isn't quite right for us," weren't always just weasel words used by editors trying to be polite. He quickly lost track of the number of occasions when, having just finished reading some manuscript, he realized that those very words described what he'd just read, and sometimes he could hardly elaborate on it. It really can happen that a piece of writing, even if very good in its own right, just doesn't fit for reasons that even the editor can't fully articulate. It was as if after reading many hundreds of submitted manuscripts, and rejecting almost all of them, he found himself sensing which rejected ones were very close to right but still had to be rejected for reasons that were operating mostly in his subconscious. What to say? Not quite in the spirit of what we do here? A little outside our realm? It might not speak well of his ability to do his job as an editor, but it's how it was, and his colleagues felt the same.

So don't be surprised if occasionally you can't figure out what you might have done wrong; maybe the reason you got rejected was the editor's mistake, not yours. That's especially true if you're dealing with one of the new online fly-fishing magazines that seem to come and go like the morning hatch. These magazines are often started by entrepreneurs with little or no experience in writing or publishing, and sometimes it shows—not only in the magazine's content, but also in the way the editor deals with writers. That's because he or she may not yet be familiar with the long-established "rules" of the game as we've explained them here. Don't avoid online magazines for that reason, however; some are excellent, even if they are the products of inexperienced "editors," and their fees buy just as many groceries as those paid by old, established print magazines.

But if, after all this, your queries don't get results no matter what you try, just do what you would if you were out fishing: Change flies. Only in this case, try a larger one.

Sending the Whole Manuscript

While many magazines are happier receiving queries, some prefer to receive a complete manuscript and others will consider either. So if you've finished your story and polished it to perfection, and your target magazine's submission requirements (if you can find them) indicate it's one that will consider a complete story, then by all means let them see the whole thing, along with a brief covering note or letter. It's OK to mention your previous publishing credits, if any, but keep it short and let your manuscript speak for itself, trusting it will prove more convincing than a query. Just make sure you give the magazine your email address so it can respond.

Your chances of receiving an acceptance are always enhanced if you provide an array of good color photos along with your story. This is where things can get a little technical because most magazines—online as well as print—have their own requirements for photo submissions. If the submission requirements are posted online they will usually spell out photo requirements in detail; if not, you can always send an email and ask for them. Some magazines have separate photo editors, so again check the submission requirements or the magazine itself; if it has a photo editor, send your email to that person.

Most magazines now prefer to receive digital photos, usually either TIFF or JPEG files. If you submit photos with your manuscript, be sure each image has your name and contact information (and the photographer's name if someone different from yourself) and a concisely written caption of the scene displayed in the photograph.

As with queries, you can make simultaneous submissions—sending your story and photos to more than one magazine at a time—but it's not a good idea. Some publications will not even consider simultaneous submissions. And how would you feel if one magazine accepted your story for, say, $300, and you agreed, and then a week later you received an acceptance from another publication that offered $500? It's almost always better to send your package to one magazine at a time and wait until you hear from that one before you try another.

How long should you wait? Who knows? The only sure thing about response times is that they vary, and as with queries, you might get a quick response or you might have to wait weeks—or, sometimes, even months. But if your patience expires before you receive a response, it's probably time to try another magazine.

When the glorious day finally comes and you receive word that Magazine X wants to buy your manuscript, the next question is to determine exactly what rights the magazine wants to buy.

Reading Your Rights

It might surprise you to learn that it's not really your story the magazine wants to buy; it's the right to publish the story. This used to be a pretty simple transaction—the magazine would offer to buy "first serial publication rights," which gave it the right to be first to publish your manuscript as a single article or divided into a series. Often the magazine also would ask for geographical exclusivity, such as "first North American serial rights," meaning it wanted to purchase the right to be first in the United States and Canada to publish your story. That type of agreement leaves you free to sell remaining publication rights in other parts of the world. In either case, after first publication, all rights would remain with you.

But that was then, and this is now. As the world has become ever more digitized and complicated, the sale of publishing rights has followed suit. Now, if you're dealing with an online magazine, it will ask for the right to publish your story online, perhaps without specifying the maximum number of times it can publish it or without defining an area within which it may do so. If the magazine has both print and online editions, it will probably ask to buy the first serial publication rights *and* the right to publish online, likely without any geographical exclusivity. In addition, some may ask for the right to publish your story, or excerpts of it, in promotional materials or in other media.

And there you are, basking in the glow of your first magazine sale, when all of a sudden you're confronted with a confusing series of choices concerning matters about which you probably have little knowledge. Of course you can always consult a lawyer, but that would probably cost more than the magazine is offering to pay for your story.

So what to do?

Our advice: *Be careful!* The magazine will probably send you what amounts to a contract, either in print or by email, and ask you to sign it. Read it very carefully until you're certain you understand all the terms. The key thing to remember is you want the agreement to define and limit the number of times the magazine can publish your story, and the number of different ways it can publish it, while preserving your ownership of all future publication rights after the magazine has used those you've granted it. Some magazines, either purposely or because their editors don't understand these things very well, may ask you to sign an agreement that, in effect, gives them *all* rights to your story for all future time. Unless you're willing to sell all the rights, don't do that! Your story might turn out to have future value you may not even have considered.

If you have questions about the meaning of any terms in a publishing agreement, ask the magazine editor as nicely as possible to explain what it is you don't understand. The same applies if you think something needs to be *added* to the agreement. The rapid evolution of online magazines has raised many questions that have yet to be answered in the law; for example, if you sell first-publication rights to an online magazine, does that also allow subscribers of the magazine to download your material and make it available to others? If you're concerned about that, or similar issues, ask to amend the publishing agreement with some restrictions.

If the answers you receive from the magazine seem wrong, or if it refuses your proposed amendments to a publishing agreement, then, hard as it may be, you should consider withdrawing your manuscript and photos. Never sign a publication agreement that restricts your own future use of the material or contains any terms you don't fully understand.

Information about publishing rights is not easy to find, but some websites provide it. Writers' associations such as the Authors Guild, National Writers Union and Outdoor Writers Association of America also can offer assistance, at least to their members. But there's a catch: In most cases you have to be a published author to join them.

Working with Editors

Magazine editors are not widely known for good bedside manners, but that's not part of their job description; at least we never saw it in any of the hiring paperwork we filled out for our various editing jobs. If they want you to change something, they usually have good reasons for asking you to do so, even if they aren't obvious to you. Regardless of your editor's age or disposition, you should always try to remain on his or her good side and stay that way.

Even the grouchiest editors can be helpful. As a young newspaper reporter, Steve had to adjust quickly to having his stories read and corrected by copy editors. Sometimes, if they found him in some especially egregious goof, they'd hustle across the newsroom to lecture him in front of other reporters, which was always humiliating. That got Steve in the habit of studying his stories as they appeared in the newspaper to see what the editors had done to them, which helped him eliminate future errors. Nobody is immune from mistakes.

That experience also made Steve develop a reluctant appreciation for what editors do, keeping writers' embarrassing mistakes from being displayed to the public. That appreciation has only increased over the years, and now neither Steve nor Paul would ever think of publishing anything not first subjected to the scrutiny of an eagle-eyed editor; neither should you. It's also a very good idea to get in the habit of reading the published version of your stuff to see what the editor did to it; that will help you avoid future errors, as it did for Steve. Editors are the last line of defense against making yourself look stupid, or—worse yet—maybe even exposing yourself to an accusation of libel.

Steve and Paul aren't the only ones who learned to appreciate editors. Lefty Kreh, certainly one of the most famous and prolific fly-fishing writers of the last generation, acknowledged in his autobiography that his writing success was due to editors who cleaned up his often-tortured prose. Steve especially appreciated his honesty because he edited Lefty's work for two magazines, and it was a real challenge. Lefty always seemed in a hurry—there were all those fish to catch!—evidently never read over his material before sending it in, and sometimes it was virtually impossible to figure out what he was trying to say without consulting him by phone or email for clarification.

Lefty wasn't the only "great" fly-fishing writer who lacked the educational background or polish to write well—Lee Wulff was another whose prose needed a lot of propping up—and many editors wouldn't have had the time or patience to work with such writers if they didn't nearly always have something important or worthwhile to say. So the moral of the story is that you should always have something important or worthwhile to say, even if you don't know exactly how to say it in easily understandable or proper English. Lefty and Lee, both famously great anglers and teachers, were more than worth any extra effort their writing might have required; until you're in that rarified league, never presume to behave that way.

Here's a fundamental bit of advice about how you should view being edited: Welcome it! Though we've both had rare experiences with neophyte editors who made a lot of off-base "corrections" in manuscripts, our experience has usually been the opposite.

Professional editors, especially those termed copy editors or line editors (as opposed to the Boss Editor who is really more a manager and isn't the one who will do the fine-tuning of your text) are a rare breed. They are advanced technicians of the language. They understand it as few writers do. They operate on levels of awareness of all sorts of subtle elements of English prose that even after many years of publishing you still might not grasp as well as they do. Don't waste your time getting defensive about their markups. Be grateful.

Even more important, learn from what they do. Any time an editor introduces you to some subtlety of the English language, they're offering you a chance to improve, so take it. Subsequent editors should never have to tell you about it again.

Years ago, Paul was helping a young writer develop his basic writing skills. This extremely hardworking fellow had overcome tremendous cultural disadvantages to reach the point where he was publishable, and Paul was glad to go over his manuscripts in painstaking detail, making extensive marginal notes that explained why this or that grammatical construction could be improved, and so on. But after doing that to three or four manuscripts, Paul noticed that he was correcting and explaining the same mistakes again each time. The young writer was just going through Paul's edits, blankly incorporating the suggested improvements, and sending them off to be published. He missed the whole point of the exercise, which was to learn from the improvements so that he'd never need them explained to him again. Don't be that way. If an editor makes a change you don't understand, just ask, without sounding defensive, why the change was made. You will almost certainly never learn enough that you no longer need editing, but you'll make everyone's work, including your own, a lot easier.

But it will make things a lot easier both for you *and* your editor if you know how to write well.

If you want to read a perfect book, there is only one way: write it.

—Ambrose Bierce

CHAPTER EIGHT

Writing Fly-Fishing Books

The difference between writing a story for a fly-fishing magazine and writing a fly-fishing book can be measured in months or, more probably, years. It's also likely not all that time will be spent on writing; depending on the type of book, much or even most of the time might be spent on research. In other words, writing a book requires a major commitment of time, effort, discipline and persistence. It's a big deal, a demanding and extended project that makes magazine stories pale by comparison.

We've already devoted chapter 6 to addressing the often-asked question of what to write about. Fly-fishing literature as it stands today is all the proof you need that we're not going to run out of topics. The number of books written on subjects about which previous books have already been written goes to show that fly fishers never get tired of those subjects, and reveals a special interest in books offering a new twist to an old subject. And we will never grow tired of a good story.

In this chapter we'll discuss where book ideas come from and offer a variety of what might be called "big-picture" ideas. In the next chapter we'll dig deeper and go through the makeup of a book from front to back, discussing the various elements that make the whole thing as complete as it should be.

Before we get started, however, it's worth reemphasizing what we said back on page 83 about artificial intelligence: Don't succumb to the temptation to try using AI as a shortcut to writing a book; it's not only unethical but will probably just get you in trouble and very likely destroy your reputation as a writer. Any book you write should be the exclusive product of your own mind.

Writing a Book Worth Reading

Starting in the early 1970s, when fly fishing experienced the dramatic renaissance and increase in popularity that we've mentioned before, the reading public's appetite for expertise likewise increased. In the wake of the fabulous success of Doug Swisher and Carl Richard's landmark book *Selective Trout*, which has now sold more than 200,000 copies, books on all aspects of the subject flooded the market. None would sell as well as *Selective Trout*, but some were genuinely inspired, many were competent and quite helpful, and some were, well, not so much. It seemed as if everyone wanted to be a famous fly-fishing writer.

As much fun as all this new reading matter was—and we both read it avidly—and as enthusiastically as many of us waited for the next crop of new books, it must be said that all this sudden verbiage coming from so many unfamiliar sources eventually raised both eyebrows and questions—not only about just what is an expert, but also about how are we to know who to listen to or trust.

This was unfamiliar ground for older fly fishers, who were accustomed to seeing new books appear at a slow and manageable pace and almost always with modest to poor sales. Several of the 20th-century books we now regard as classics, including *Art Flick's Streamside Guide to Naturals and Their Imitations* (1947), and Vincent Marinaro's *A Modern Dry-Fly Code* (1950), had poor sales in their first editions and weren't widely appreciated until reprinted in the commercial prop wash of the *Selective Trout* publishing boom.

Seventy years ago, when Flick, Marinaro, and others were quietly doing their original studies and creating their pathbreaking flies, whoever could have imagined there would ever be sufficient public appetite for speakers among the fishing clubs to keep so many experts busy on the fly-fishing lecture circuit? Whoever imagined there would even *be* a fly-fishing lecture circuit, much less that a quirky and comparatively obscure sport such as fly fishing could ever generate actual celebrities? But there they were, and whatever else we thought of them, their very existence alerted us that fly fishing was now something very like a community, with common values, commerce, passions, and an unbridled enthusiasm to communicate with one another. The sport would never be the same.

All that being said, and keeping in mind Lefty Kreh's famous definition of an expert as "a guy more than 50 miles from home with his own slide projector," there really are authentic experts, and we really do need them. We need them in the many businesses that provide us with gear, and we need those experienced, insightful individuals who, through their articles, lectures, videos, blogs, websites and books help us get better at being fly fishers—not just at catching fish but at protecting fisheries resources while learning the sport's essential values and complex heritage.

Whether you intend it or not, when you start publishing you unavoidably become, at least in the public eye, one of those experts. We both have experienced the awkwardness of finding ourselves perceived as knowledgeable or skilled, sometimes far beyond what we really were, simply because we were authors of fly-fishing books. We both like to think that we were, in fact, entitled and qualified to write what we wrote, but we never intended to set ourselves up as omniscient all-wise authorities on anything. Actually, being regarded that way gives us the creeps.

What's New, and What's Not

There are good reasons for offering this warning, the most important being to remind you that it's all too easy to start believing your own promotion and get careless. This is especially true if you think you've got a new idea and you're eager to share it in an article or book. All too many authors have rushed a book into print only to have readers and (even worse) reviewers point out there's really nothing new in it, that in fact the basics of your idea were laid out in some other recent article or in a book published long ago. How embarrassing.

Don't misunderstand this point. It's perfectly OK to revive and celebrate worthy old ideas if it seems like they're being neglected, or if, for another example, you are in fact writing a manual that explores a variety of historic fly styles or historic fishing techniques. Any time you can help your fellow fly fishers broaden their range of skills and tools, you're writing something worth publishing. But when you do so, it's only fair, and good manners, to let your readers know you're not claiming to have originated all of this stuff. In our first few years fishing we are most likely to "discover" or "invent" something that's only new to us but old to other fly fishers.

Look at it this way: Two hundred years ago (more or less; such dates are always approximate), fly fishers, mostly British at the time, were gradually abandoning horsehair in favor of silkworm gut as the best possible material for their terminal tackle. One hundred and seventy years ago, some adventurous British anglers were tying large gaudy streamers to catch coastal saltwater fish from small open boats. One hundred and fifty years ago, in both Europe and North America, anglers who for generations beyond counting had relied entirely on rods that were essentially just solid wood sticks found themselves adjusting to the exciting fishing opportunities provided by split-cane rods of much more complex construction. Eighty years ago, the quick popularization of nylon leader material was a huge breakthrough, genuine news of the first magnitude for anglers whose parents and grandparents had relied on the weaker, less reliable silkworm gut, thus replacing a former innovation with a new one. Fifty years ago, we were all eager to read the first field reports of the then experimental graphite rods. None of those developments happened instantly. There were die-hard British fly fishers still using greenheart rods well into the 1900s—there probably still are today—and early nylon's limitations kept many anglers using gut leaders at least some of the time until the 1960s. But all those are examples of genuinely new things in fly fishing.

Many other seemingly new ideas weren't. Flies that swam through the water upside down, as do many modern weedless and "keel" patterns, were specifically recommended and illustrated in Robert Venables' *The Experienced Angler*, published in 1662. Flies that imitate mice, frogs, and other non-insects seem to have been around since anglers noticed that big fish love such foods. The modern popularization of hinged-bodied flies is just a rerun of similar patterns described intermittently for more than a century. The same with long trolling flies that have a second hook trailing on a section of leader a short distance behind the primary hook; we've seen this style fly illustrated in British books from the early 1800s on, and it has popped up, or been reinvented and fancied up, again and again ever since. Even the deservedly acclaimed Swisher-Richards book *Selective Trout*, for all its perceptive insights, contained a number of effective old-idea revivals, including spent-wing spinners, split-tail dry flies, and both no-hackle dry flies and emergers. This isn't to take anything away from Carl and Doug; their great and lasting contribution was to reinvent, revive, and even refine those things so we'd never again forget them. But it is a further reminder of how little under the sun is actually new in fly-tying styles, even if almost everybody has forgotten a given style's last incarnation decades or even centuries ago. It's also another great reason to read the older books.

The truth is that once you get outside the technological realms in which some new piece of gear is actually making something easy or even possible that didn't used to be—such as a previously unknown rod or line material, an array of synthetic materials for flies, or waders that don't leak, and such—there's little in fishing that is wholly new. What there is instead is an often very exciting succession of small modifications and improvements of old ideas. The savvy writer will constantly watch that succession of modification and refinement for story ideas.

A wonderful advantage of being the sort of well-read fly fisher we encouraged you to become back in chapter 2 is that you'll be exposed to all those historic ideas in their earlier forms. You'll be in a position to track a good idea through its various permutations, to notice the theoretical dead ends of that process, and to become all the more aware of how many generations of fishermen before you thought their way through the same problems you face. If you're paying close attention, studying those previous incarnations of good ideas can lead you to fruitful writing of your own. Earlier, in chapter 2, we took a harsh view of telling the same old stories again and again, but the rediscovery of old ideas, handled right, is always fodder for new writing.

A remarkable example of the revival of old ideas in a new book is the inspiring story of the late Sylvester Nemes, who, starting with his first book, *The Soft-Hackled Fly* (1975), revived and celebrated a then-underappreciated style of fly. His several later books elaborated on the history and advantages of such flies, and led to a nationwide popularization of what are also known as "north-country flies" because in the minds of many anglers in the UK and the United States they are most closely associated with the streams of Northern England and Scotland. Today, thanks to what Nemes started and then other writers and tackle companies explored further, practically every American fly shop you visit these days will have a generous selection of traditional and modern variations on the soft-hackled fly.

But what takes Syl's accomplishment even further into the rarefied domain of a historic fly-fishing achievement is that eventually his books also led to a similarly grand revival of this style of fly in the UK. This was a rare if not unique case of an American fly-fishing writer restoring an important element of the British fly-fishing tradition for the very people who had developed it in the first place but had largely lost sight of it. Imagine accomplishing something that far-reaching and historic.

Where Do Books Come From?

Books come from good ideas, which in turn can come from unexpected sources. When Steve was young his father taught him to fish with flies for Kamloops trout in British Columbia, but as he grew older Steve wanted to learn more about these fish. Except for a couple of chapters in Roderick Haig-Brown's classic *The Western Angler* (1939), he was surprised to find virtually nothing in the popular print, so he began hanging out in libraries, especially the library at the University of Washington School of Fisheries, where he had access as an alum. There, searching scientific journals, he discovered many research papers containing a wealth of information about Kamloops trout and their environment, including entomology, geology, limnology and botany.

That was back in the 1960s when photocopying technology was in its infancy. Copy machines were expensive to use and not always available anyway; the technology also was still primitive and sometimes produced copies that quickly faded away, so Steve had to copy some articles by hand or check out journals and type the information they contained. It was a long, labor-intensive process, but eventually he assembled enough information about Kamloops trout to fill several three-ring notebooks. Recognizing the information he had gathered was potentially of great value to anglers, Steve realized it would be selfish just to keep it to himself, and that's what led to the idea of a book.

But to make it complete, he needed a lot more information. That meant visiting and fishing a large number of Kamloops trout waters (a classic case of a tough job, but someone had to do it), interviewing fisheries managers and hatchery workers, visiting fishing camps from posh to primitive (with emphasis on the latter) and learning their customs and how they were managed. At the posh ones, this might have meant learning when one was expected to wear jacket and tie at dinner; at the primitive ones it meant discovering whether you needed to bring your own pots and pans, sleeping bag and kerosene lantern. Steve also had to track down local fly patterns, learn how to tie them and trace their histories.

It was a huge project and it took several years to complete, but a lot of it also was fun—Steve got in lots of fishing, saw some marvelous country and made some lifelong friends—and it was a memorable moment in 1971 when he was finally able to hold the first published copy of the book in his hands. The original book subsequently went through two fully revised editions over a 30-year life span, and those old three-ring notebooks are now part of special collections at Western Washington University in Bellingham, Steve's hometown.

As Steve says in his account of writing and occasionally revising *Kamloops*, we often write books because we want to learn the subject, or at least fill in gaps in our own knowledge. Another way of saying that is that the book we end up writing is the book we wish we'd had when we first became interested in the subject. Remembering how little you once knew about a subject can keep you sympathetic to the reader's lack of knowledge so you won't leave out any vital information.

Paul's first fly-fishing book, *American Fly Fishing: A History* (1987), was one he didn't even realize he'd needed to read (had it existed) when he started fly fishing. From 1972 to 1977 he spent much of his time working as a ranger-naturalist and park historian in Yellowstone National Park, where he found the park's natural and human history an irresistible subject, and by 1979 had begun publishing what over the next 40-some years turned out to be a long series of books relating to the park.

As a historian by training (his MA thesis, 1976, was a study of Yellowstone's historic archives), he naturally gravitated toward the park's history, which led him deep into the literature of the American conservation movement. Because sportsmen have always been key components of wildlife conservation, he was necessarily exposed to the huge parallel literature of American outdoor sports, which gave him a significant leg up when, in 1977, he left Yellowstone to become first executive director of what was then called the Museum of American Fly Fishing in Manchester, Vermont (it's now called the American Museum of Fly Fishing, the better to embrace the global scope of its collections). He arrived at the museum knowing how much more he had to learn, but at least he had a solid historical framework for that learning.

During his years (1977-1982) at the museum, Paul was first approached by a publisher about writing a history of the sport, though even at that point the idea of writing such a book didn't originate with him. In the mysterious and often inexplicable way of these things, that first offer fizzled out, but a few years later, after he'd left the museum and was living in Montana writing full-time, an arrangement was worked out between the publisher Nick Lyons, the museum, and Paul, and he finally did write it.

By then Paul had been studying fly-fishing history for a decade and voraciously reading the sport's historic literature even longer. Five years as editor of the museum's journal, *The American*

Fly Fisher, had further acquainted him with a wealth of the latest historical research and, perhaps even more important, with many of its leading authorities (this was how he first met Steve, who wrote some timely articles on Pacific Northwest fly fishing for the journal). As further unintentional preparation, and thanks to the seasonal nature of his Yellowstone employment and grad school terms before coming to the museum, Paul had vagabonded the country, fishing many waters from Florida to Washington and California to Maine, which gave him at least some acquaintance not only with a great diversity of historic fly-fishing waters but with the regional and local traditions that had grown up around each.

Still, much like Steve with *Kamloops*, even after all that research and experience Paul realized that before writing the book there was more study to be done, including a trip back to Vermont to paw through the museum's growing collection of treasures, then on to Worcester, Massachusetts, to get a look at some of the rarest early American sporting periodicals at the American Antiquarian Society, then on to Yale for a quick look at a few items in their fabulous Beinecke Library. Then he came home and wrote the book.

That was hardly the end of the matter, though. Like many who love writing books, he soon discovered that books cause more books. Eventually *American Fly Fishing* became the foundation of a series of more specialized books that allowed Paul to explore all sorts of obscure corners in the sport's rich traditions and modern practices. In books of essays, personal narratives, and even fiction, he continued to discover the inexhaustibility of fly-fishing writing.

These later books offered additional examples of how books happen. Most notably, from 2003 to 2009 Paul wrote a history column for *American Angler* magazine. This allowed him to investigate many historical topics he had dealt with only briefly in *American Fly Fishing*. His experiences writing a couple dozen non-fly-fishing books trained Paul to perceive any article he wrote as potential book material (recycling is good), and these magazine columns were no exception. But magazine columns are typically quite short, only 500 to 1,200 words, so Paul first wrote a full-length essay on each chosen topic, often 3,000 to 4,000 words, then trimmed it down to column length for the magazine. Material from the longer versions ended up as large parts of two books and the entirety of a third.

There's a lesson we've learned in these processes. While some of our books were written from beginning to end all in one extended period, others were in part or entirely collections of previously published periodical material. It's important to practice this sort of literary economy. Get the most out of your words as is reasonable and fair. This week's story or essay might with suitable revision become part of a chapter in next year's book. Thinking this way will also provide you with an opportunity to check your own past thinking; you will be surprised at how often you'll want to revise some previously published piece before using it in a book.

IDEAS THAT SUCCEEDED

History and biography are both growing in popularity among fly fishers. Has your favorite river ever been given a good biography? There are many excellent models for that kind of book, including the late John Merwin's *The Battenkill: An Intimate Portrait of a Great Trout River* (1993); Ed van Put's *The Beaverkill: The History of a River and Its People* (1996), or the collaboration of many writers and others to produce *The Creel: North Umpqua River Edition* (2008). Don't imagine that just because your local stream may be less famous than these that it's in any way unworthy of a well-researched and thoughtful written chronicle of its anglers and their stories.

The same holds true for articles about the sport's graybeards, whose adventures and contributions to the sport are bookworthy. Models of thoughtful and informative angling biographies of notable fishermen include Jack Samson's *Lee Wulff* (1995); Eric Leiser's *The Dettes: A Catskill Legend* (1992); Graydon Hilyard's

Carrie Stevens: Maker of Rangeley Favorite Trout and Salmon Flies (2000), and Valerie Haig-Brown's dual biography of her parents, *Deep Currents: Roderick and Ann Haig-Brown* (1998).

We can always use more creative natural history writing in our fishing books. Many of our best writers have been superb naturalists; they have to be terrific observers of the ecological setting to understand the streams they fish. For just one outstanding example, Paul has often described British nymph-fishing pioneer G. E. M. Skues' masterpiece, *The Way of a Trout with a Fly* (1921), as natural history thinly disguised as a fishing book.

As Paul's experience indicates, fly-fishing history in all its forms can inspire endless different book ideas. One example is the late Harry Middleton's acclaimed memoir, *The Earth Is Enough: Growing Up in a World of Trout and Old Men* (1989), which celebrates his boyhood years in the Ozarks learning to fish and much more from his uncle and his grandfather.

A much different view of fly-fishing history can be found in Kirk Wallace Johnson's 2018 book, *The Feather Thief: Beauty, Obsession and the Natural History Heist of the Century*, which created quite a stir. It's the account of a brazen thief who stole many priceless rare bird skins and used them to tie classic Atlantic salmon flies or sold them to other fly tyers. It was a great idea for a book, combining fly tying with real suspense and outright crime, but let's hope we won't be seeing another on the same subject.

Thanks to the experiences of new generations of anglers, changes in the fish's world, and science-driven changes in our understanding of the fish, there will always be room for another book about brook trout, or cutthroat trout, or striped bass, or any of the other fish we enjoy catching. Maybe you'll write one of them.

For some years, at least a few anglers have taken on the challenge of trying to catch every species, subspecies, or strain of a given fish, usually a trout. For just one example, Nick Karas' *Brook Trout: A Thorough Look at North America's Great Native Trout*

(1997), shows just how wide and deep the research trails are in presenting a literary portrait of a single species. And that being said, there were any number of earlier books about brook trout, and there have been more since. Books like Karas' provide many ideas and models for the writer shopping for subjects to pursue further.

Rather than focus on an exhaustive historical and scientific portrait, journalist M. R. Montgomery, in *Many Rivers to Cross* (1995), chose to document his quest to fly fish for the many rare and threatened subspecies of cutthroat trout on a fly. Maybe it's time for someone else to find out if those rare trout are still around, or to focus in on the native fish of a smaller region or even of a single river.

It's also worth keeping an eye on the taxonomists who, now advantaged by DNA-related science, occasionally rearrange some long-familiar genealogy of fish we thought we knew but now have to reacquaint ourselves with as they've jumped to another branch of the family tree. There's always a story there, whether it's just to report on the change or to check with other taxonomists who disagree with the new classifications.

As we said in chapter 6, many writers have tried writing about casting and only a very few have been successful. By the late 19th century, expert fly-fishing writers were publishing diagrams trying to show the desired flight of a properly cast fly line, and some were pretty successful at it, but it wasn't really until modern photographers got involved using cameras fast enough to stop the motion of the line that various fly casts could be accurately portrayed on paper. Two of the most influential modern books on this subject are Lefty Kreh's milestone *Fly Casting with Lefty Kreh* (1974), the first casting manual to take full advantage of such photography, and *The Essence of Fly Casting* by Mel Krieger (1987), which had not only great photography but design and production values worthy of a fine-art book. But as good as those two books are, they reinforce our point that fly casting is amazingly idiosyncratic, each of us finding our way to what works best for us. Lefty and Mel had raised fly casting to its highest and, frankly, most beautiful form, but they taught dramatically dissimilar casting styles, proving, as if it needed to be proved, that there are even more ways to cast a fly than there are to skin the proverbial cat.

These two classics have been followed by a host of equally sophisticated productions by a variety of excellent casters—Joan Wulff, Ed Jaworowski, Tom Deck, McCauley Lord, to name only a few who have devoted entire books to various aspects of the subject—to the point that, though there will always be a need and a market for casting manuals, new books on the topic have long ago reached a point of diminishing returns as far as offering much new information.

That's not to claim there is nothing new to say. Fly casting is like every other element of the sport; there is always more to talk about, if only as the technical advances in tackle manufacturing open up new possibilities for specialty casting. Lefty himself returned to written instruction of this evolving skill set in 2008 with publication of his massive and in many ways comprehensive *Casting with Lefty Kreh*, an oversize, full-color, 456-page photographic extravaganza that improves on his original casting book in many ways. In fact, if you're considering writing about casting, perhaps the first thing you should do is spend some time with Lefty's enormous and authoritative book to see just what's been said already.

Fish Stories

Many fly-fishing books are based on their authors' personal experiences, and some of these have become great public favorites. Examples include Roderick Haig-Brown's great personal memoir, *A River Never Sleeps*, as well as his famous "seasons" series—*Fisherman's Winter*, *Fisherman's Spring*, *Fisherman's Summer* and *Fisherman's Fall*, as well as Robert Travers' ever popular *Trout Madness* and *Trout Magic*, or Russell Chatham's *Silent Seasons* and *The Angler's Coast*, or Tom McGuane's *The Longest Silence*, or W. D.

Wetherell's *Vermont River*, *Upland Stream* and *One River More*, or Steve's books *The Year of the Angler*, *The Year of the Trout* and *Rivers of the Heart*. And we could go on and on.

Any one or more of these could serve as a blueprint for someone who wants to write about his or her own fishing experiences. While many contain useful information about fly fishing, that really isn't the point of any of them; you might think of them as campfire stories in print, told just for the entertainment of their audiences. They are the kind of books you want to curl up with on a long winter evening, or use as a thoughtful gift for a fishing friend in the hospital or otherwise temporarily confined for reasons of health. They also offer opportunities to try different writing styles and demonstrate the full lyrical or emotional depth and power of your prose, what some people call "the music of writing."

Books of this type are the best argument for why you should keep a detailed personal diary of your fishing exploits; it will give you lots of grist for your writing mill. And don't forget to keep a tablet and pen next to your bed so you can jot down the inspirations and ideas that sometimes come only between midnight dreams.

Such books also often give you a chance to recycle material you've published before, perhaps as magazine articles that may fit comfortably as chapters in a series of essays about your fishing encounters and adventures. Otherwise, there is no particular formula for writing about your own experiences, except to suggest you write the kind of stories you'd want to hear or read yourself.

Use Your Imagination

So far we've had little to say about fly-fishing fiction. Of course, some people will tell you that *all* writing about fly fishing is fiction; that's because anglers historically seem to have gained a reputation for exaggerating the number and size of the fish they catch. But you don't have to worry about things like that when you're writing fiction, and that's one of its charms. It's also more fun to write fiction than any other type of book; one writer has described it as like "coloring outside the lines," and that's a good analogy. There are hardly any rules in fiction, and the borders of your imagination are the only limits.

Both Steve and Paul have experience writing fiction. Besides three non-fishing novels, Paul's "Shupton's Fancy," a tale of the rediscovery of a 15th-century fly pattern that never failed, first appeared as a small, stand-alone book, then as a selection in his later book of fictional short stories, *A Fish Come True.* Steve is author of two book-length collections of fictional short stories, *Trout Quintet* and *Six Fish Limit.* The former included "One Size Fits All," about a 12-year-old boy who buys a battered old fishing hat at a thrift store and discovers it has almost magical properties. The latter book included "The Fishlexic," the story of the world's first genetically modified fly fisherman, and "The Man in Black Waders," a novella-length yarn about the trial of the world's most famous fly-fishing writer, who was being sued for plagiarism.

Where do such ideas come from? Mostly just from imagination, or from something we've heard or experienced in the past. In Steve's case, years ago he was fishing with an attorney who told him about a client who was trying to convince the Internal Revenue Service that the expenses of a fishing trip should be tax-deductible. At the time it struck Steve as a good idea for a story, but it was nearly 30 years before he figured out a way to write it. It first appeared in *Gray's Sporting Journal* and was later revised to become the first selection in *Six Fish Limit.*

So although it's true most publishers don't appear anxious to see them, we've managed to prove that fictional fly-fishing books *can* be published. Now it's your turn.

Subjects to Avoid

Before we leave this subject, we should mention again, as we did in chapter 6, that there are some subjects you should probably avoid, including fictional storylines or "plots" that have become hopelessly shopworn. This is nothing unusual, and it's hardly limited to fishing writing. Plenty of writing manuals will tell you about the five—or six, or seven, or whatever small number—basic fictional plot lines that prevail in all novels. Whether such generalizations about the art of novel writing are gross oversimplifications or not, it's still true that it all comes down to what the novelist makes of his chosen plot. This is as true in writing about fly fishing as it is for any other subject.

Novelist William Humphrey, quoted earlier on bad writing, and best known for such novels as *Home from the Hill* (a National Book Award winner for 1958, leading to a 1960 movie starring Robert Mitchum, Eleanor Parker, and George Peppard) and other acclaimed works of fiction and scholarship, was an enthusiastic outdoorsman. When he turned his hand to the very often used plot of the giant-old-trout-that-couldn't-be-caught he produced a fine tale, *My Moby Dick* (1978), which no one has regarded as boring or hackneyed. There's plenty of life in some of those old chestnuts if you have the skill to find it.

Humphrey's little book also demonstrates there are sometimes good reasons why we tend to tell versions of the same stories again and again, and it's not just because writers are too lazy to think up something new. Serendipitous streamside encounters with huge fish and strangers will always be a part of our sport and will always generate a lot of adventures and misadventures. Among the best tales of meeting another angler with unexpected consequences was written by another famous novelist-angler, Robert Traver (*Anatomy of a Murder*, 1958, leading to a 1959 movie starring Jimmy Stewart, Lee Remick, and George C. Scott). We'd also hate to have to get along without the powerful story of such an encounter as Traver's "The Intruder," in his book *Trout Madness* (1960).

All this is to say we realize some things are going to keep happening on streams, and each generation of anglers rightly wants to share its own version of those things; after all, it's the first time it's happened to them, and for many readers it will be the first time they've heard of it, whatever it is. Your job isn't as simple as just avoiding all these old storylines; it's to recognize not just their limitations but whatever possibilities might still be wrung from them, then seeing if you can offer a fresh angle.

Take Your Choice

By now you probably have in mind a good idea of the book you want to write, whether fiction or nonfiction, and you're anxious to get started. But there's still an important decision to be made: Will you write the entire book before you try selling it to a publisher, or prepare and send a detailed proposal of your book idea to a potential publisher and hope the reply will include a book contract?

There are pros and cons to each approach. If you decide to write the entire book and then try to sell it, you're gambling that a great deal of time and effort might go to waste if your book doesn't sell. By contrast, a book proposal—although it also requires a substantial investment of time and effort—involves much less work than writing the whole thing. However, writing the book first also can result in substantial benefits; the lengthy and laborious process of committing your words to printed or digital form often helps your book take on something of a life of its own, leading to a sharper focus than you originally had in mind. When that happens, you may end up with a more saleable book. Either way, you'll still need to prepare an elaborate, complete, well-researched, eye-catching proposal to submit to one or more publishers; we'll describe how to do that in chapter 10.

Paul and Steve have done it both ways. Depending on the type of book it is, Steve prefers to write the whole book first and then submit it to a publisher. That's especially true if you're writing fiction; in that case, you pretty much *have* to write the book first, unless your résumé includes enough previously published fictional works to earn a publisher's trust that your next work will be worthy of consideration. But if you're writing a nonfiction book dealing with one of the many aspects of fly fishing, Steve's experience has been that writing the entire book first helps clarify his concept of the book. Paul agreed with that approach most of the time, but on occasions when he wanted to get an advance on a book project that was going to take a lot of time, he followed the common procedure in that case, of writing two or three sample chapters to be submitted along with a summary and outline of the whole book, so the publisher would have a reasonable idea of what he had in mind. By the way, in Paul's experience once the publisher was confident enough to sign him up, they didn't really mind if the sample chapters had to be changed a lot. That often happens as the writing continues and the best way to organize the material evolves.

So those are your choices, although there are some examples of exceptional circumstances where you might not have to write the book first or submit a book proposal. It doesn't happen often, but if you're already an established writer and/or fly fisher, a publisher might call and ask you to write a book, which has happened to Paul and Steve. It also usually helps if you're really famous; for example, former President Jimmy Carter never had to worry about finding a publisher for anything he wrote about fly fishing or any other subject.

Most writers aren't that lucky, however, and must choose between the alternate approaches described above. If that includes you, then give each alternative careful thought and decide which will work best for you. Then write for the ages.

And further, by these, my son, be admonished: of making many books there is no end; and much study is a weariness of the flesh.

—Solomon, Ecclesiastes 12:12

CHAPTER NINE

Anatomy of a Fly-Fishing Book

When you're writing your book and later when you're discussing it with editors, designers or anyone else involved in its production, there are at least a few things you need to know about the traditions and terms of bookmaking. It's a fascinating subject in its own right, especially the convoluted history and craft of the making of the physical book, whether it's an ancient, one-of-a-kind book handwritten on parchment, or the product of a modern printing technology that is still evolving more than five centuries since the first books came off a printing press. But we'll stick to a few basics here and leave it up to you how much further you want to pursue that subject. This chapter describes the well-established conventions for how the parts of a book are organized between its covers.

First a bit of terminology. Choose a book from your shelves and open it flat on your desk, and you will see a "two-page spread." The page on the left side is the verso, and the one on the right is the recto (lame jokes uncalled for; printers have heard them all). Page one always appears on the recto, which is to say that the right-hand pages will be odd-numbered all the way through the book.

The contents of a book have three main parts: the front matter, the text, and the back matter. We're going to work our way through the main parts and their elements. Some we will discuss at length, others need little or no explaining here. Just so we don't have to repeat it again and again, we will emphasize right off that you will find an almost infinite number of variations on this "standard" form, but for most purposes this is how it works:

Part One: Front Matter

The front matter is numbered separately from the text that follows it. Rather than using the "regular" Arabic numbers we're accustomed to, the front matter's pages are numbered in Roman numerals, starting with "i" on the first recto. Printer's preferences and the vagaries of assembling the pages may mean there is one or more completely blank page at the very front of the book, especially in older books; we have nothing to say about those here except they're a nice place to write down a phone number if you can't find another piece of paper when you need it (don't do this if the book is a rare first edition, however).

Page i, half-title: The half-title page, on that first recto, lists only the main title, not including the subtitle. This is also sometimes referred to as the "bastard title," though we don't hear that much anymore. The title is the only thing on the page, which is one big reason why many authors prefer to sign and inscribe books on that page; there's lots of space.

Page ii, blank or other: The back side, or "reverse" (thus verso) of the initial recto is often blank, but many authors use it as an opportunity to list their previous books. We find that approach useful both as writers liking to show off a little and as readers wanting to know more about the author's previous work. Alternatively, and admitting that this doesn't seem to happen as often as it used to, this first verso is a nice location for a mood-setting photo or work of art, in which case it's known as the *frontis*. There was a time when the frontis was commonly a portrait of the author, but that's a tradition we're just as glad has faded away.

Page iii, title: The title page typically includes the full title and subtitle, followed beneath by the author(s)' name, perhaps followed by the name of other involved parties, such as the illustrator(s). If needed, this is also where readers are told a little more about the book's history, such as that this is the "third revised edition," or is "compiled posthumously from the author's letters by K. Trout." Then, toward the bottom of the page, the publisher's name and location appear. Some publishers used to put the date there too, but not so much anymore.

If you're designing and publishing the book yourself, please notice that the author's name is all the title page needs. Don't say "by Rod Snapt" or, even worse, "written by Rod Snapt." Just your name.

Many authors prefer to sign and inscribe their books on the title page, though their decision to do so may depend in part upon how much empty space there is.

Speaking of titles, what's yours going to be? Good question. In 1930, Pulitzer Prize–winning biographer, poet, and folklorist Odell Shepard published one of trout fishing's great literary gems, a small book—more an extended essay, really—called *Thy Rod and Thy Creel*. The print run was small, and the book quickly disappeared from view, only to be discovered and brought back into print 50 years later thanks to the combined enthusiasm of three of fly-fishing literature's modern masters: William Humphrey, Arnold Gingrich, and Nick Lyons, who published it. As Arnold said, "If ever there was a book of our century that I'd like to see commended to the anglers of the next, this would have to be it." Nick's paperback reprint of the book, published in 1984, did make it available to thoughtful readers, but like its predecessor this new edition also received little attention. The book remains unknown among most angling readers. Have you ever heard of it?

Why should such a splendid book fare so poorly in the marketplace? The consensus of opinion is that the fault lies with the title, which, being so apparently ecclesiastical in tone, put off many potential readers who assumed it was more about religious matters than fishing. But the book is no such thing; it is an eloquent yet easygoing 120-page rumination on trout and trout fishing, composed almost entirely of such memorable passages that, as Arnold said, "it is almost a disservice to quote from it" because "it should be swallowed whole." Sadly, it stands as angling literature's foremost example of the way really bad things can happen to a book if its title stinks.

As with titling articles, coming up with a good book title is a tricky craft. Shepard's book notwithstanding, it's rare that a title will completely ruin a book's chances of finding a reasonable audience. Even if the author gets the title catastrophically wrong, a conscientious editor is likely to step in and negotiate something a little better. But it's a sure thing that some book shoppers who, when they see a book with a mundane but basically functional title, will reflexively relegate it to what fishing writer Craig Woods once called "the pit of sameness" instead of wondering if it might have some new lessons to offer, or fun stories to tell. This is especially to the disadvantage of an author whose name the shopper doesn't already know; we all tend to have more interest in books by a writer we already know and trust.

It is much easier to title the how-to books. People shopping for them expect and possibly even prefer a nuts-and-bolts title along the lines of Charles Brooks' *Nymph Fishing for Larger Trout* (1976), John Merwin's *The New American Trout Fishing* (1997), or Tom Rosenbauer's *The Orvis Guide to Beginning Fly Fishing* (2009), which make it plain what they're going to tell you. Hundreds of helpful and well-received fishing manuals have proven that you can hardly go wrong just saying what the book is about in the simplest way possible.

Still, there's a lot to be said for giving your instructional book a title with a little more flair. Mel Krieger's *The Essence of Fly Casting* (1987) and Datus Proper's *What the Trout Said* (1982), suggest by their titles that maybe these books will offer not only instruction but some added literary flair, or even inspiration, which both do.

It's the books with ambitions beyond straightforward instruction that offer bigger challenges and bigger rewards for coming up with engaging or intriguing titles. Call them what you like—literary, experiential, reflective, or just storytelling—these books, especially if you're a new author, are a bigger gamble for the publisher because, as they keep telling us, books of fishing stories just don't sell as well as fishing manuals. The exceptions to this vague rule are spectacular, though, and it has been our good fortune that we've always been able to find publishers willing to give our books of stories a try. Maybe they thought we had superior audience recognition because we'd published so many other books. Or maybe they dreamed that whatever book we'd offered them would be one of those rare ones that takes off, sells like crazy (or just sells slowly but for a very long time), and eventually even gets made into a movie. Naturally, we are happy to let them dream such dreams, if that's what it takes to get them to publish the thing.

One proven way to come up with a title for your book of essays or stories is to extract a lively, telling phrase either from your own book or from some other literary work that you invoke in your text. Stellar examples of this approach include such classics as G. E. M. Skues' *The Way of a Trout with a Fly* (1921) and Norman Maclean's *A River Runs Through It* (1976). A variation on this approach is to use the title of a chapter that appears in the book, such as Nick Lyons' *Fishing Widows* (1974), Ernest Schwiebert's *Death of a Riverkeeper* (1980), and Thomas McGuane's *The Longest Silence* (1999).

But examples of good writers cleverly summing up their book in a title of a few intriguing or otherwise appealing words are legion: Robert Travers' *Trout Madness* (1960) and *Trout Magic* (1974), Howard Walden II's *Upstream and Down* (1938), Roderick Haig-Brown's *A River Never Sleeps* (1947), Harry Plunkett Greene's *Where the Bright Waters Meet* (1924), John Gierach's *Trout Bum* (1986, and in fact all his other cleverly titled books since that one), Margot Page's *Little Rivers* (1995), and, ever and always, Walton's *The Compleat Angler* (1653), which sums it all up, whatever *it* is, in those three words.

We've both encountered publishers who were reluctant to use titles they found too vague because someone in their marketing department was nervous about a title that didn't lay out with some precision what the book was really *about*. Though there is good reason to respect their concern—why no one stopped Odell Shepard from using such a misbegotten title is an enduring mystery—all you have to do is invoke *A River Runs Through It* or *The Longest Silence* to show just how freestyle or whimsical a fishing book's title can be. But a further defense of such titles is provided by adding a straightforward, explanatory subtitle. This is done all the time. Datus Proper's *What the Trout Said* had a subtitle that continued and clarified his intentions: *About the Design of Flies and Other Mysteries*. McGuane's subtitle was *A Life in Fishing*. On the other hand (and to return to beating a dead trout), as if to nail the lid on the coffin of his book's potential sales, Shepard's *Thy Rod and Thy Creel* didn't even have a subtitle. Ironically, all but the first few pages of his book were internally titled "Thread of the River," which, while perhaps not great, would have been a huge improvement on his chosen title.

Choosing a title for your book, whether simply explanatory or ethereally poetic, should be fun. Enjoy it. We give our titles a lot of thought, and may go through interim "working titles" before arriving at our choice. Paul is currently nearing the completion of a book of stories whose folder is still titled *Another Fishing Book.* And don't forget that you're not alone in this important quest for just the right title. Many editors, who have seen countless books displaying many titling philosophies cross their desks, might have just the right idea for you, or might at least serve as a sounding board for your deliberations. Don't hesitate to seek such help; you're entitled.

Page iv, copyright: This page is a catchall for a lot of vitally important information, including not only the book's copyright, ISBN numbers (International Standard Book Number; look up this essential ingredient on your own), edition number and other printing details, and a host of other specifics. It's a handy place to put lots of things, such as a statement of the book's adherence to archival paper standards; or certain brief types of acknowledgments; to additional information about the author; to credits for illustrations (including cover artwork) or specific quotes from other authors that appear in the book. As we will show later, most of these things can be put elsewhere, and probably should be. You should work with your editor to assure that the copyright page is attended to thoroughly. There's important legal stuff going on there.

We add only two further bits of advice on the copyright page. First, this is an opportunity for us to repeat an earlier warning that some publishers have boilerplate (first-draft default language) contracts that simply assign the copyright of your book to the publisher. Unless that's part of a clearly discussed deal that gets you some pretty big returns, absolutely do not let them have the copyright. It's your book, your work. The book should be copyrighted in your name, and this page should say so. Among the many reasons you want to hold the copyright is that if—and this actually happens—the publisher abruptly goes out of business next year, the rights to your book won't be tangled up indefinitely in the lawyer-abundant hassles of sorting out the corporate properties.

Second, more generally, being a good reader means knowing as much as possible about a book and its author. As an alert reader, whenever you start a new book you should get in the habit of reading this fine print; you may find yourself surprised by how useful it can be to your eventual take on the rest of the book.

Your book provides you with a rare opportunity, one matched in few other professions, to express, publicly and permanently, your admiration, appreciation, or love to someone very important to you, or even to their memory. You can make that expression not only to a person or persons, but to any other entity that stands high in your esteem and that will be visibly honored right there in the front pages of a book a lot of people will see. We recommend not missing this chance.

Page v, dedication: Book dedications have a long and entertaining history. In Walton's day, for example, authors (including Walton) routinely dedicated their books to wealthy or otherwise influential people, composing flowery, pages-long dedicatory statements in praise of said worthy. Many of these were genuine heartfelt tributes to friends, some we today suspect to have been otherwise motivated, but they shared a windiness that, thank heavens, often didn't continue into the actual text. It's just how it was done then. Here's an example (spelling modernized, capitalization retained), a single paragraph that was in fact milder than many, excerpted from Richard Brookes' four-page dedication of *The Art of Angling* (1740) to a friend:

> It is not my Purpose to offend your Modesty by going into the usual Style of Dedications, however I may just mention what will not lay me under the least Imputation of Flattery,

> That You are the Delight of all Companions where you happen to be, and are possessed of every Quality that constitutes a Complete Gentleman. This naturally puts me in mind of many agreeable Moments which I have lost by my Removal from your Neighborhood, and which I cannot think of without Regret, especially since I learned from you the serener Pleasures of Life, and to enjoy all the Advantages of a rural Retirement.

In those days of such courtly talk it must have been heart-warming to the recipient of a few pages of this sort of gushy thing when it was sincerely offered, but it appears that extreme dedicatory groveling was also a good way to earn the favor of some personage or patron, especially a wealthy one, who either was not insensitive to the social cachét the dedication might carry, or who just loved wallowing in flattery. However fine or unsavory those authors' reasons, it must have worked well at least some of the time, because authors kept doing it. After all, if your dedication did happen to hit the bullseye in Viscount Bigcastle's ego zone, and he invited you to spend a few comfortable weeks or months at his country estate bordering a rarely fished salmon river, a few pages of adoring prose was a small price to pay.

Luckily for us, modern dedications are much more succinct. Almost all are brief, along the lines of "for Alma" or "to the boys at the club" or "to the good works of Suckers Unlimited." Though this is all entirely up to you, and some modern dedications are longer and still quite satisfying, remember that you will have a better opportunity to expand your expressions of gratitude in the acknowledgments (see below).

Also, remember that there is no need to say "This book is dedicated to . . ." A simple "To" or "For" or "In memory of" or something else concise will do.

Just for variety, here's a rare exception to such brevity. In an approach very few of us would take but probably wouldn't fault—and we're not implying that just anyone could get away with this—Ernest Schwiebert went long in dedicating *Death of a Riverkeeper* to the memory of an esteemed friend this way:

Arnold Gingrich
(1903–1976)
Faithful flyfisher and many-sided man of letters who loved books and fine writing, honored the lessons of Walton and Halford and Skues, worshipped music and its sister arts, fished with a mixture of skill and elegance and love, savored literature and the humanities and fine cuisine, committed himself totally to preserving our bright rivers tumbling swiftly toward the salt, heard the magic in their music, gloried in the quicksilver poetry of the fish themselves, and happily collected many things over many years—fishing tackle, paintings and sporting prints, books, antique motor cars, toys and miniature artillery pieces, elegant split-cane rods, exquisite vintages and rare violins, and a richly varied and celebrated circle of friends.

One last tip about dedications, and it's an important one: A surprising number of publishers may try to bury your dedication on the top of the copyright page, where it is immediately lost in the welter of ISBN numbers, copyright statements, and other fine-print administrivia. This bewildering behavior may be some misguidedly cheap effort on their part to try to save a little paper by mashing these two front-matter ingredients together, or it may just be the result of a problem we discuss for different reasons elsewhere, that some book designers view all text not as made up of individual words that say things but as "design elements" that can be shuffled about almost interchangeably without affecting the book's functions. Whatever the cause, don't let them diminish the impact of your dedication by squeezing it into the wrong place. Insist—repeatedly if you're not sure they're paying attention—that your dedication must appear on a page of its own—the recto of the two-page spread. Properly placed there, the dedication gives you plenty of room to depart from your customary practice of signing books on the half-title or title page, and write a more fulsome personal inscription to the dedicatee right under the printed dedication.

Other Front Matter

From this point on, the pagination becomes more complicated because, beginning with the dedication, additional pieces of the front matter are optional and might not even be included.

You may include an epigraph; that is, a quote from some notable figure whose words sum up something important about the intent or spirit of your book. Some authors favor quoting two or three such pithy wisdoms to more fully suggest the flavor of what's coming. This quote might even contain the phrase you used for the title. The best way to learn how to employ an epigraph is by reading a lot of them.

The table of contents is not dispensable; it must be there. But it and other subsequent prefatory items, such as a list of the book's illustrations and maps, or a list of tables, may begin on either verso or recto. The table of contents is just what it says it is. Lists of illustrations or of tables are rarely needed in fly-fishing books, though we've seen a few cases where they might come in handy.

Depending upon how you, your editor, your book's designer, and your publisher's style sheet see it, you may want to have any or all of the following: a foreword, a preface, acknowledgements, and an introduction. If for some reason you decide you have to have them all, and in a moment we'll advise against doing so, that's the sequence in which the four should appear, each preferably beginning on a verso.

Foreword: Traditionally a foreword is written by someone else on behalf of you and your book but lately we're seeing many authors writing their own "forewords" when what they are really writing are prefaces.

And another word to self-publishers: your misspelling "foreword" as "forward" will be the reader's first clue this is not a professionally published book.

Will a foreword help sell your book? Should your book have one? The answer to both questions might be yes, but before we get into the reasons why, let's first examine the difference between introductions, prefaces and forewords. A preface is an author's way of introducing him- or herself and explaining why and/or how the book was written and what the author hopes it will accomplish. For that limited purpose it also helps promote the book, but not as much as an introduction or a foreword might. An introduction also is usually written by the book's author and provides a sort of road map of what's inside the book; it also helps promote the book, more than a preface does but probably not as much as a foreword might.

From a promotional standpoint, a foreword is the next step up from dust-jacket blurbs. Typically the foreword is written by a widely known and authoritative person whose deep familiarity with your book's subject, or even with you as an individual, entitles them to speak at greater length, usually one to three pages of text at the front of the book. The foreword writer's substantial credentials can give your book a big leg up in gaining the confidence of prospective buyers. Forewords are promotional fanfare, announcing that the book is so worthy of your attention that it merits its own up-front ceremony, pages in which a distinguished figure holds forth on why the book is such a great thing. In this, forewords serve as stage-setters, often explaining why this new book is significant in the context of the books that have preceded it. Notable examples of that type of foreword in fly-fishing books include one written by the late Joe Brooks, at the time the universally admired grand old man of trout fishing, for Swisher and Richard's highly successful *Selective Trout* (1971), or one written by Vincent Marinaro, a recognized leading theorist in fly-pattern design, for Datus Proper's *What the Trout Said* (1982). In both these cases, the senior figures were providing more than a fluffy little blurb; they were placing the new book in the greater context of its area of expertise.

The style and approach of forewords varies depending upon the book. A book of literary essays might best be served by a foreword from a well-known essayist, even a literary figure of note from beyond the bounds of fishing writing. A book of angling entomology might benefit most from a foreword by a notable university scholar of entomology. It's all about elevating the shopper's first impression of your book as one worth having and reading.

But a foreword might not always be necessary; most books get along fine without them. As far as promotional needs go, several dust-jacket blurbs might do the job best. Most bookstore shoppers seem unwilling to stand and read a book's entire foreword before deciding whether to buy it. It's not a life decision they're making, after all.

Online buyers, on the other hand, especially those considering a book with a "Look Inside" button that allows them actually to read some portion of the book, might be a little more likely to sit and read the whole foreword. But we suspect most shoppers who are looking for trustworthy opinions and recommendations about a book that interests them will be content with some persuasive blurbs. In fact, for them, just the existence of the foreword, written by a writer whose name they recognize, might be all they need.

Which leaves only a less tangible but perhaps still important reason for having a foreword. There is personal and professional satisfaction for you in having such concrete evidence that a distinguished authority thought your book worth the trouble of writing a foreword for it. Also, it's a distinction that adds to your credibility when it comes time to submit future manuscripts for publication.

So if you're interested in having a foreword, talk it over with your publisher. If they publish a number of other fly-fishing writers, they may know which of them are most approachable. A responsible foreword writer is doing you quite a favor when the most they can expect for their work is a warm feeling of having been helpful, and maybe a complimentary copy of the book.

Preface and acknowledgments: The preface is usually a brief, almost pre-introductory statement explaining or justifying why you wrote the book. We prefer that any acknowledgments should be placed under their own heading as part of the book's back matter, but you will often see acknowledgments of one sort or another variously scattered here and there, maybe on the copyright page, in the foreword or preface, or in the introduction. It's usually best for all purposes to consolidate your thank-you's under one heading, and we will have more to say about that below, in the discussion of the back matter.

Introduction: we're bucking tradition just a bit here by suggesting that it makes more sense in most books to regard the introduction as a part of the text rather than as part of the front matter, so we will cover it in our discussion of the text,

There are several other pieces of the book's superstructure that often wind up in the front, all of which add additional explanation of something in the text to come, such as a list of unusual citation abbreviations, or a list of main characters in a novel, or a translator's notes. We don't need to go into them here, as they almost never apply to fishing books.

If we have a single overarching guideline for front matter, it is to get through it respectful of its function but with a minimum of ceremony. Let the reader get on with being a reader. Put as much of the book's professional armature in the back matter as possible.

Part Two: The Text

Though the text is the heart and almost all the body of a book, we are going to say comparatively little about it. Almost everything we've said so far in this book leads you here.

Even assuming, as we do, that all the previous chapters will be a big help to you, it's safe to say you still have a lot of learning ahead of you in the creation of your book's text. Developing the skills needed to create a sensible and easily followed narrative for 40, 50, 60, or more thousands of words is in some unavoidable respects a matter of on-the-job training.

We all go about organizing such a long discussion in our own ways, with our own approaches to making outlines and tentative tables of contents. Even if you just start on page one and wing it all the way through, you're going to be relying on a great deal of thinking you've done before you started. This process is most straightforward in the writing of instructional books, whose lessons usually follow in some intuitively obvious order.

It can be harder to come up with the "best" sequence of material in a book of stories that are really connected to one another only by the author's voice, but neither you nor your publisher should worry too much about that; the fact is that an author's dedicated readers probably don't care in the least which story comes first or last; they just want to read more stories from that person.

What follows, then, is an accumulation of useful tips that reach beyond these sweeping generalizations.

Page numbers: This is as good a place as any to point out, in case you haven't noticed in your own reading, that a book's page numbers are sometimes left off certain pages. This is a design feature employed on pages where the page number is regarded as unnecessary or, perhaps more important, is an actual distraction from the page's purpose. This is why you'll likely not see a page number on a title or half-title page, or on a page devoted entirely to an illustration, whether artwork or a photograph. It's a longstanding design nicety to let that page enjoy its one primary purpose without the intrusion of a little number in the corner.

Introduction: When the first segment of the text is titled "Introduction" it does just what you'd expect an introduction to do, but it introduces more than just the subject; it also introduces the author's "voice" and establishes the general tone of the text, whether cordial or cold, conversational or formal, discursive or disciplined. It also may serve the purpose of a preface, as described above, if you haven't included one. Many fly-fishing book introductions also try to set a mood, say with a couple of anecdotes. Others summarize in order every chapter in the book; this approach follows the old saw mentioned earlier in the context of writing articles, "First tell 'em what you're going to say, then say it, then tell 'em what you said," but it's often helpful in giving readers a sense of what's coming.

Introductions are more often short than long. Whether it's a book of technical instruction, lofty philosophical essays, or zany misadventures, it can probably benefit from a few hospitable opening words, the author's way of welcoming the reader into the book.

Chapters: Dividing the book into chapters serves more than the obvious purpose of sorting what you're going to say into tidy, manageable chunks of information.

Chapters provide important guideposts to the reader's navigation of the book, but you should also think of chapters as a kindness to the readers. More chapters aren't always better than fewer chapters, but an author should sympathize with the reader's patience and try not to let any single chapter drag on too long. Just what constitutes "too long" is not always easy to determine, but think of your own reading experiences, and the times when your foremost feeling as you finished a long chapter was relief rather than enthusiasm. Don't make your readers feel that way. If a chapter has that much material in it, it's probably time to rethink it, maybe even to break it into two chapters.

There's always room to experiment with how you put your material together in a succession of chapters, but there are reasons why certain conventions have arisen and endured over time. For centuries writers of all-around fishing instruction books have tended to start with such basics as the tackle and other gear to use; these discussions are especially intriguing to read in the earliest books, especially those in which the author was justifiably obligated first to teach readers how to build a rod from a single whole stick of just the right wood—the right species and size, selected, straightened and cured under just the right conditions—or to find just the right stallion, convince it to share some of its tail, and then weave those hairs into a functioning fly line. Modern writers go through similar introductory exercises to explain the differences between monofilament and fluorocarbon leaders, or the countless styles and weights of fly lines. Writers of intermediate to advanced instructional books will often shorten or just skip these discussions, assuming most of their readers already know enough to move on to whatever new information the author proposes to offer. There is no one right way to do all this, any more than there is one right way to organize your book of fishing stories.

Within the confines of a single chapter, many writers, both instructional and other, have found it helpful to introduce subtitles that break the subject into more manageable chunks. These "sub-chapters" reflect smaller transitions in your narrative than those requiring a whole new chapter. They also are good for preventing reader fatigue, especially if you're deep into a discussion that is necessarily dry or intensely detailed.

It's usually not necessary to break your book into two or more parts (each containing several chapters), but it's done now and then. Be open to the possibility if you think you can do it in a way that helps the reader navigate your book's subjects.

Photos and illustrations: So far we've said little about the use of photographs or illustrations, but these are often critical components of fly-fishing books. Remember the old saying that a picture is worth a thousand words? Well, sometimes it's worth even more. Can you imagine a book about fly tying that lacks photos or illustrations of the patterns being described? Or one about casting without photos of casting sequences? Or a book about tying fishing knots without illustrations of the knots? Of course not. And if you're writing about a particular fishing venue, it's best not to leave it up to the reader to imagine what the place looks like; provide a photo or two. If that place also happens to be hard to reach, you need to provide at least one good map that shows the way. Even if none of those things apply, and your book consists of a series of essays on fishing, a good artist can provide illustrations that will bring some of those essays to life, or emphasize whatever points the essays are trying to make.

If the illustrations—let's say they're pen-and-ink or pencil drawings—are ornamental rather than instructional (that is, if they're included primarily to evoke the mood of a story), they're most often placed at the top of the first page of each chapter above the chapter title. But you should pay attention to whether or not that placement is the most effective for the purposes of your book. If the drawings in question have a lot of fine detail, they can stand to be much larger. If so, it may be best to give them more space, elevating them from a tiny decorative bit above the chapter title to a prominent feature of the book. For example, you might decide that it's best to devote the entire verso of the opening spread of each chapter to the drawing while the recto contains only the chapter title and the opening paragraphs of the text. There's no question that the larger the image, the better able the reader will be to enjoy it properly.

Be forewarned: the editor, designer or someone else in the publisher's office might bristle a bit at the added pages and resulting increase in paper costs of giving the illustrations greater space. And it's true that the possible good effects of such a change are impossible to measure in dollars and cents. But making the most of the artist's work is just a good idea. Also, keep in mind that giving the illustrations greater attention could require other adjustments. The images have gone from being a bit of pretty trim to being a major feature of the book. Should more be said about them in promotional material? Should the artist be more prominently featured on the book's cover (as in "Illustrations by Homer Winslow")? Your editor will be your best advisor on that.

Where do you get such photos or illustrations? It helps if you happen to be married to an accomplished artist, as Paul is to Marsha Karle, who has illustrated a number of his books. But if you're not that fortunate, you can still make friends with artists, photographers or cartographers, and it will help a lot if you get to know them well and spend some time fishing with them. That will help them understand you and what you're trying to accomplish.

One of Steve's closest fishing friends, Alan Pratt, chief editorial cartoonist for the *Seattle Times*, provided two beautiful paintings of fly patterns and several excellent maps for Steve's first book. Other artistic fly-fishing friends, including August Kristoferson, Dave Whitlock and Al Hassall, created illustrations for several of Steve's other books. It fulfilled a longtime dream of Paul's when his book *Royal Coachman* had illustrations by the celebrated artist Eldridge Hardie.

As for photos, many fly-fishing authors prefer to shoot their own, but there also are professionals who specialize in the exacting work of photographing such difficult subjects as casting sequences, fly patterns or natural insects. If you don't know any, just search some recent books featuring those subjects and you'll probably find the names of several. Your publisher also may maintain a list of professional photographers, illustrators and cartographers accustomed to working with authors.

Of course, you'll have to pay these people for their work. You should consult with them in advance to come up with an agreement that sets forth in detail exactly what you expect from them, how much you will pay them, and whether payment will be in installments or on final acceptance of their work. Your agreement also should specify when the work is due—some artists and photographers are not accustomed to working on deadline. The amount you agree to pay should depend on what work you want done, and good artists or photographers often command hefty fees. That means you will have to calculate how much the fee may subtract from the advance or royalties you expect to receive for your book. In other words, it's a little bit of a gamble on your part, and you must decide whether it's worth it. And speaking of royalties, unless the book is a true partnership in which the author and illustrator have shared in its creation—in which case they will share the royalties—we recommend against arrangements in which, in lieu of advance payment, the illustrator is given a cut of the royalties. A gifted illustrator should be paid well and generously; their images can make a big difference in the effectiveness and success of your book. But it's still your book and the royalties should likewise be yours.

Sometimes publishers will decide a book needs illustrations, photos or maps not provided by the author, and commission them on their own initiative. If that happens, you can be sure the publisher will adjust your book contract accordingly, reducing the amount of the advance and/or royalties to offset the cost of the illustrations. In such a case it is vitally important that you are ensured adequate opportunities to review the resulting illustrations and retain absolute rights of rejection over them.

So, unless you're a gifted or experienced artist or photographer yourself (or happen to be married to one), illustrations in any form will cost you some money. But never underestimate the importance or utility of good illustrations in your book; just think of them as an investment in its success.

But you should also be forewarned: When you start submitting photos or illustrations to your publisher, it greatly increases chances that something will go wrong. Steve regretfully recalls the exasperation he felt when he opened the published copy of his first book and discovered the publisher had paired several photo captions with the wrong photos. The photos all had been submitted to the publisher with their captions physically attached, but just in case any of the captions became detached, each photo also was keyed by number to its matching caption. It seemed impossible anything could go wrong—but it did. What made the errors especially obvious to readers was that one caption mentioned snow on the ground in the photo, but that caption ended up on a photo where it was quite evident there was no snow. An editor who was paying any attention at all should have seen that at a glance and fixed it, and although Steve had received and corrected proofs of the book's text, he never saw any proofs of the photos and captions, and by the time he discovered the errors in his book it was far too late to do anything about them. The mistakes were eliminated in two subsequent fully revised editions of the book, but Steve still feels the bitter disappointment he felt from his first look inside the published book. Paul has had similar experiences both with fishing books and with books on other subjects. We can't overemphasize the extent to which the people who design your book, bless their well-intentioned hearts, can and will be breathtakingly oblivious to what the captions say and the pictures show. We could share other stories of such problems, but the point is made. Besides, it upsets us too much to talk about it.

The moral of the story is that if your book has photos, illustrations, maps or diagrams, make it absolutely clear to your editor that you *insist* on being the last person who sees the proofs of the book, *including all illustrations and captions*, just to make absolutely certain the publisher and editor have got them right before the book goes to press, That means you will need to work very closely with your book editor, copy editor and proofreaders to assure everything is correct.

That same precaution should apply even if your book is text only, with no illustrations or photos. We know of one book where a proofreader made many unauthorized changes to the final proofs, which were never sent to the author; some were grossly erroneous or altered the meaning of whole paragraphs of the text. Yet another book, written in two distinctly different parts, had separate introductions to each part, but the publisher unaccountably placed both introductions at the front of the book, which not only confused readers but left the book's second part more or less adrift without explanation. Still another book, with many color photos whose tiniest but most important details were explained in great detail in the captions, fell into the hands of an expletive-worthy technician at the printer who cheerfully and in utter ignorance of the harm he was causing, photoshopped the essential details right out of the photographs, making the captions sound delusional. These examples of neglect on the part of publishers, editors, proofreaders and almost anyone else with even the slightest chance to gum up the works, underscore the absolute necessity to make sure your eyes are the last ones to see the proofs before the book goes to press—especially if it has photos or illustrations.

But if you take the necessary steps to escape such errors, illustrations, photos, maps, and diagrams can add substantially to the informative value of your book and its attractiveness to readers.

Illustration credits: You have options with these, and your editor will most likely have preferences for how credits should be handled. If each illustration's source is already credited at the conclusion of the caption, you've fulfilled your obligation and nothing else is necessary except in extraordinary circumstances, such as when you have received a lot of help from a particular photo archive or other institution in the research and locating of illustrations (e.g., historic photos or other artwork that may not be generally available). In those cases don't hesitate to thank them again in the acknowledgments. Except in the case of a book with illustrations by one or more persons throughout, in which case the illustrator(s) might be listed on the title page, most photo and illustration credits in fly-fishing books are handled in the captions. If the photographer or illustrator or other source of artwork has specific requirements for crediting their work, they will let you know; do what they ask. It's usually only in illustration intensive books, such as coffee-table books, or "photo essays," that all credits are bunched in a separate section in the back of the book.

Conclusion: Most instructional books and many noninstructional books offer some sort of wrapping up—a final discussion that may, indeed, "tell 'em what you said," but can serve a number of other purposes. It's usually toward the end of the book that many authors include a plea for stream conservation or for good manners on our ever more crowded rivers. The most forward-looking authors might even use the conclusion to suggest where readers interested in continuing study of their book's topic might best go, which is a nice way of admitting that, well, yes, I've spent 30 years learning what I've just told you, but I am not vain or delusional enough to think there won't always be more to learn or even that I've got it all right.

Epilogue: The conclusion is the proper end of a book. If there is no formally designated conclusion, then an epilogue serves the same purpose. If there *is* a conclusion, then an epilogue may offer post-book thoughts, often of an "Oh, by the way" sort. It's a place to put afterthoughts, or mention events or facts that came to your attention after the book was written, or that wouldn't necessarily have merited attention in the book but are of sufficient interest to bring up afterward.

Afterword: Afterwords can function as wrap-ups much as do conclusions and epilogues, but they sometimes appear even in the presence of conclusions and/or epilogues. Or, in subsequent editions of the book, the publisher and author might agree that an "afterword" is in order. Afterwords thus offer an opportunity to report on new developments relating to the book's subject, or news about things that have happened or have been learned since the book was published. In that role, afterwords serve the purpose of adding to the book's topic without going through the time-consuming and more expensive process of updating the various parts of the text and creating a completely revised edition of the book. A few books are by nature amenable to easy addition of chapters. Bob DeMott's *Angling Days: A Fly Fisher's Journals*, which used excerpts from his many years of fishing journals to tell the story of his fishing life, originally appeared in 2016; three years later, the trade paperback featured several additional journal entries from 2016 to 2019 without the need to modify anything in the first edition. However, after enough time has passed, and enough new information has accumulated, an afterword becomes unwieldy. If you believe the situation has reached this stage, then it's best to try to persuade the publisher that it's time to revise the whole thing. We've both done that with books, and we recommend it. Besides allowing you to keep the book up to date, you get the satisfaction of knowing that the book attracted enough of an audience, and enough validation from the fly-fishing community, to be worth a new life in an updated edition. Congratulations; that's at least as worth celebrating as the initial publication of the book.

Part Three: Back Matter

The back matter of almost all fly-fishing books is brief. As with the front matter, there may be several potential elements of the back matter, such as a glossary of terms or a list of abbreviations, but most are applicable only to more technical works. Each back-matter element will start on a verso.

Acknowledgments: It's probably true that acknowledgments more often appear in the front of books than in the back matter, but we prefer them to come after a person has read the whole text. It could be argued, though, that by placing the acknowledgements in the front matter you are showing the reader the extent of your study and research, thus further establishing your credibility and increasing the reader's confidence in you. Though that's certainly true in some kinds of books, we doubt that fly-fishing book readers need that much cheerleading to keep them reading. The back of the book is soon enough for them to learn how you did your homework, if they care at all. As we've said, we prefer the front of the book to have the least clutter possible. As always, your editor may disagree, and it's not such a big deal that you should argue over it.

This is not to say acknowledgments don't matter. They are important, and though we urge you to avoid making them sound like an Academy Award acceptance speech ("I'd like to thank all the little people, starting with my third-grade teacher Mrs. McPherson, who taught me not to chew on my eraser"), make sure you honor your sources of information, advice, and even inspiration.

The acknowledgments also are a good place to make note of any previous publication of portions of the book, most often sections of the text that have already appeared as magazine articles.

It's good manners to thank those magazines, even to single out individual editors who were of special help in preparing the original articles. If you've significantly revised an article's texts for use in your book, it's appropriate to point that out. Some authors lean toward a succinct statement: "Portions of this book previously appeared in an earlier form in *Martian Canal Fly Fishing*, *Trophy Sculpin*, and *Floss Quarterly*." On the other extreme, some authors specify which chapters were derived wholly or in part from which articles, and may even give complete formal scholarly citations for each article. This is a good thing to discuss with your editor, to see if there's a style sheet for it. In whatever way you choose to handle this, just make sure you do give those magazines credit. They helped get you to this point. Salute them.

Otherwise, thank away. Check how other authors have handled their acknowledgments. Fly-fishing book acknowledgments can afford to be lighthearted, and often include lists of fishing friends who shared the experiences or studies that led to the book. Credit is sometimes given for timely sandwiches or a few nights' floor space. More professionally, if a tackle shop or manufacturer was kind enough to share gear or important advice, it's only fair to thank them. Thank your book's editor, or thank the publisher's entire editorial and design staff who, after all, transformed your raw digital file into a handsome book.

The sequence of all these expressions of gratitude is up to you unless otherwise instructed by your editor, so you just have to decide how you'd like to prioritize them.

Last, if your book is full of strong opinions or adventurous theoretical thinking, it's a good idea to conclude the acknowledgments with a statement absolving all these helpers of blame, to the effect of "I thank all these many friends, mentors, and advisors, but of course any mistakes in the book are mine." Own it.

Appendixes: An appendix is a handy place to put helpful information that lends itself to a list or other tabular form but would only be an encumbrance in the main text. It may be a list of tackle dealers that carry the necessary gear or materials prescribed in your book, or a list of fly fishing–related conservation groups, or state management agencies. Many people find such lists useful, so if you have one or more in mind don't hesitate to use them. Most books get along without them, perhaps partly because they so quickly become out of date. Businesses come and go, phone numbers change, websites die.

Appendixes are also good places for discussions perhaps too peripheral to include in the main text but are still meaningful or original enough that the reader will probably appreciate them. In that way they are similar to endnotes that elaborate upon some comparatively narrow topic but aren't essential to the main narrative. Such marginal information might be good conversational fodder for you and your editor, who can help you decide if it should be in the main text, in an appendix, or just left out of the book entirely (if the latter, don't forget to save it for later!).

Notes: Notes are rare in fly-fishing books but are sometimes necessary. The most important service they provide is to make it easy for readers to follow up on an interesting point made in the text, whether they just want to know more or want to check to see if the author actually did justice to what the source says. Some of Paul's books were based on considerable historical or scientific research by many people including himself, and were written in good part for an audience that might want to pursue this or that particular point. Paul finds *foot*notes (i.e., notes at the foot of each page) distracting in his own reading, so he relegates all the citations to the *end*notes at the back of the book, which in some cases have been quite numerous (his popularly written book *Cowboy Trout: Western Fly Fishing As If It Mattered* [2006], had a 227-page text followed by 35 pages of endnotes). Check with your editors to see if they prefer a certain style for notes.

Many people are intimidated by the mere presence of notes; somehow those little superscript numbers here and there in the text make them think the book is going to be dull and technical.

We don't know what to do for people who have fallen for that stereotype, but maybe if your designer makes the numbers as small and unobtrusive as possible, shy readers won't notice them, or at least won't mind them as much.

Bibliography: Many readers love suggestions for more reading but few fly-fishing books include bibliographies, or list only a handful of titles that are especially relevant to the book's topic, in which case they just can be titled "Further Reading." More extended bibliographies are handled in various ways; again, the publisher probably has a house style you'll use.

Bibliographies are always alphabetical, by the author's name. The simplest bibliography will present reasonably full citations (author, title, publisher, place and year published) for the books or other publications mentioned in the text. The next step up is to include those titles plus any others that have been important to you in your work. Another common approach is to present the bibliographical information in a narrative style, a "bibliographical essay," as a chapter-by-chapter summary of what books and other publications were most helpful in the writing of each chapter.

Index: Most fly-fishing books get along fine without an index, but if your book is one you hope readers will refer back to for information on specific points you've made, then you should consider having one. The simplest indexes include only proper names (places, people, book titles, and a few other items), and will be useless to someone wanting to locate and reread your discussion of upstream emerger fishing techniques. If you're going to have an index, think about how your readers are most likely to use it and make sure you're including what they'll need.

The most ambitious attempt at comprehensive indexing we've seen in a fly-fishing book was in Ray Bergman's milestone *Trout* (1952 edition in this case), which concluded with no fewer than five indexes: Tackle and Equipment; Methods and Tactics; Fly-tying Tools, Methods, and Materials; Persons, Places, and Literature; and (just to be extra sure he'd covered it all) Miscellaneous.

Even admitting that *Trout* was a big book, and a great one at that, we don't believe you need more than one index if you make sure it does what your readers will want it to do.

Publishers may resist adding an index, as it does add extra pages and is perceived to be a bother. By setting it in smaller type, the designer can greatly reduce its consumption of pages, maybe even to two or three pages total. Today's indexing software eliminates almost all of the time-consuming tedium of making an index, but the results can still be glitchy and require a careful proofreading that only you are qualified to provide. Be thankful you don't have to make the index from scratch, but in a way that only old-timers are likely to appreciate, something good is lost in not having you or someone go through the word-by-word and page-by-page process of hand-indexing a book. For one thing, indexing is one of the best possible ways to review the book for surviving inconsistencies and other flaws. For another, indexing is its own art form; your first time through the process can make you painfully aware of how poorly you understand the alphabet.

Though some publishers might still be willing to foot the bill for the cost of having an index made, don't count on it. If you don't want to do it yourself (and you should realize that a professional will do a much better job, with only final proofreading oversight by you), then suggest the publisher have it done at your expense. The publisher will either know independent indexers or will have someone on staff who can handle it. You can often persuade them to take whatever it costs out of your first royalty check, so you won't be immediately out any cash. Just make sure you get to proofread it when it's done.

Dust jacket and "About the Author" copy: Many publishers will welcome a brief text—just a paragraph or two—describing your book. They can use it not only for the book's cover or the inside flaps of the dust jacket (i.e., the paper cover most new hard-bound books come in), but also for catalog copy and other promotion. Don't assume that just because your editor has read the whole thing he or she is the best person to write up a suitably concise yet glowing description of the book. Do some homework by reading other books' catalog copy and covers. This is the time to suppress your modesty and deploy your most lively, affirmative adjectives in portraying what your book accomplishes (e.g., insightful, lively, innovative, iconoclastic, original, congenial, provocative, droll, surprising, etc.). By all means stick to the truth, but don't hesitate to invoke your loftiest ambitions for what the book will bring to the reader.

Most books include brief "about the author" statements, either on the inside back flap of the dust jacket, or as the book's very last internal element, after the index, or both. The advantage of having the statement on its own immediately following the book's index is that if the book makes it into a paperback edition, there will be no dust jacket, so the internal author statement might be the only place it appears. Read other books' "about the author" statements to see how they've been handled. When writing the statement, it's good to mention publications in which your work has appeared. Other possible inclusions are relevant education (a degree in fisheries ecology would be nice, economics not so much), memberships in fly fishing–related organizations, breadth of fishing experience, and other things that might enhance your credibility. All this is hard to do if you're a first-time author with few obvious credentials, but lengthy, thoughtful experience is as good as any credential.

Like so much else in the promotion and marketing of a book, writing good promotional copy is something of a crapshoot. It's hard to predict what will work best; one very successful natural-history book Paul encountered some years ago simply said the author was a Yale graduate, that statement being, in some circles at least, a social and intellectual "credential" that probably helped sell a lot of books. Just do your best.

Next to the pleasure of reading a favourite fishing book comes that of persuading a friend to read it too.
—ARTHUR RANSOME, *THE FISHERMAN'S LIBRARY*

CHAPTER TEN

Finding a Publisher

SOMEWHERE OUT THERE IS A PUBLISHER just waiting to see your book idea or manuscript; you just have to find the right one. Don't expect it to be one of the big megapublishers we've all heard about—those that pay six-figure advances and confer instant celebrity status on the author—although a few of those giants do occasionally condescend to publish a fly-fishing book. But as we've said before, books about fly fishing are considered "niche books," which means that when you do connect with a publisher it might be one you've never heard of previously. It also might take a long, frustrating time to make that connection, but when you finally do it will change your life.

It wasn't always so difficult. When Steve and Paul started writing books, most publishers were accustomed to dealing directly with new or aspiring authors, giving their words a chance to speak for themselves. But those days are long gone, and now it's almost as if authors have been removed from the equation, even though they remain the lifeblood of the publishing industry. In these days of ever more consolidations, most large book-publishing companies accept submissions only from agents who represent authors, not from authors themselves. That includes the so-called Big Five—Penguin Random House, HarperCollins, Macmillan, Hachette and Simon & Schuster—and many of their hundreds of imprints.

How did this happen? It seems to have evolved from the notion that professional literary agents know the book business, and what types of books have the greatest sales potential, much better than authors, especially new authors. That might be so, but it also makes life much more difficult for writers, and potentially keeps big publishers from discovering talented new authors.

Does this mean you need an agent to get your book published? Not necessarily. There are other pathways to publishing, especially for "niche books." This chapter describes some of those pathways and how to follow them, considers the pros and cons of hiring an agent and how to find one, explains how to write effective book proposals, suggests ways to expand your appeal to potential publishers, lists publishers currently in the business of producing fly-fishing books, and examines some of the complexities of book contracts.

That's a lot to cover, so let's get started.

Getting Connected

It goes without saying that before you approach a publisher you must have something to offer, either a book idea, one or more completed chapters, or a completed manuscript, and the latter is almost certainly essential if you're writing fiction. Then you should do some research on the internet to see what fly-fishing books have been published recently and who published them; that should give you an idea of which publishers might be interested in your proposal. Once you've narrowed the field and confirmed that your most likely target publisher accepts unsolicited book proposals, a direct approach is best: Just send in your proposal, preferably by email if the publisher accepts proposals by that method. Later in this chapter we'll explain how to draft such a proposal, and what should be in it.

On the other hand, if you've got your heart set on connecting with one of the Big Five publishers or others that refuse to accept unsolicited submissions, you should also do a little research to identify some of the venerable publishing houses that have occasionally published fly-fishing books in the past—Alfred A. Knopf, W.W. Norton, Morrow and others—in hopes one might be willing to publish a new fly-fishing title. But you'll still need an agent to run interference for you, and we'll explain how to find one a bit later.

Other than those methods, about the only other way to connect with a publisher is if the publisher comes to you first, which doesn't happen often and mostly happens only under unusual circumstances—such as you're already a world-famous person whose name on a book is likely to make it a bestseller. A more common way this can happen is that a publisher who has already published one or more of your books, or even a different publisher who has admired something you've written, will contact you and ask you to write another book. That happened twice to Steve. The first time was when Nick Lyons, the publisher often mentioned in these pages, got in touch and asked him to write a book about steelhead fly fishing. Nick hadn't previously published anything by Steve but was familiar with his first books. Though flattered by the proposal, Steve declined because at the time he didn't think he'd yet had enough experience fishing for steelhead to write such a book. Nick was persistent, however, and over the next decade he repeatedly offered Steve a contract to write such a book. Steve, who spent much of that decade fishing for steelhead, finally agreed and the result was *Steelhead Country*, published in 1991 by Lyons & Burford and republished three years later by Sasquatch Books, a regional publisher based in Seattle.

The second time occurred when Sasquatch, after republishing *Steelhead Country* and two of Steve's other books, asked him to write a guidebook on Northwest fly fishing. That wasn't the kind of book Steve wanted to write, however, and he was involved in another book project at the time anyway, but he promised to show his new book to Sasquatch when it was finished, though he wasn't sure they would be interested because the book—*The Estuary Fly-fisher*—was aimed at a narrow audience and parts of it were quite technical. Sure enough, Sasquatch turned it down, but then, with one phone call, Steve sold the book sight unseen to Frank Amato, founder of the publishing company now called Amato Media, one of the nation's most prolific publishers of fishing books.

If this is your first book, then what happened to Steve is extremely unlikely to happen to you, although such opportunities might well come after you've established a reputation as a writer. We say "extremely unlikely" rather than "impossible" because under rare circumstances—say if you've written an important, attention-getting article on a previously unexplored subject—a publisher might ask if you've considered expanding it into a book. Or if you have a reputation in some aspect of fly fishing without writing anything a publisher might approach you with a book idea. That was what happened to Paul when he was invited to write a history of fly fishing on the basis of fairly limited credentials. Like Steve with his steelhead book, it took Paul some years before he felt up to the task, just at the same time he was conveniently approached by another publisher (this was also Nick, in fact) and the book happened. But such circumstances are rare, and even if they happen you'll still probably need to write a book proposal.

Preparing a Book Proposal

The first thing to understand about book proposals is that they are far more elaborate and complicated than writing queries for magazines; just about the only similarity is that you want your proposal to make a big impact, just as you should try to do when proposing a story to a magazine. In this case, however, the goal is to capture the interest of your target book publisher.

The proposal should be prepared in a thoroughly professional manner, typed in an easily readable font such as 12-point Times New Roman or Arial, with double- or 1.5-line spacing and one-inch margins. If you put it on paper, make sure you use only one side of each sheet.

Other than that, there seems no universal agreement on what a proposal should contain, but most require at least the following:

A brief cover letter identifying yourself and stating your purpose: to convince the publisher your book or book idea is worth publishing. Make sure the letter includes your contact information so the publisher can respond.

A description of your prior writing or publishing experience, or other relevant information, such as what qualifies you to write the book. Don't fudge.

A brief description of the proposed book or book idea, usually called an overview or synopsis. Start with the book's title and your name, then outline your story (or plot, if it's fiction), not word-for-word but with sufficient detail to give the publisher a clear understanding of what the book is about. Include a table of contents with the title and subject of each chapter to help the publisher understand how the book is structured. If relevant, include information about the possible number and type of illustrations or photographs the book will contain. Conclude with the estimated or actual length of the manuscript in number of words.

Provide a sample chapter. If you're writing nonfiction, you should select the chapter that has the greatest impact, displays your best writing or, preferably, both (some publishers want more than one chapter; just follow their guidelines). This is your chance to make your best impression. If you're writing fiction, it's usually best to select the first chapter of your book.

Market Analysis. This is probably the most difficult part of any book proposal, and perhaps the most important. It requires you to plunge into arcane matters you've probably never considered, and regardless of how much information you can dig up, your publisher—if it's a large one—will probably still know more about the sales potential of your book than you will. Yet publishers consider the market analysis a very important component of any book proposal, so it will pay dividends if you do the very best job you can.

Begin by defining the target audience for your book, which for fly-fishing books is obvious. But how many fly fishers are there in North America? You'll have trouble finding any industry association or government agency that tracks such information, and though a few websites suggest there are something like seven million fly fishers in the United States, it's hard to find any data to support that number. Even if it is correct, your target market might be only a small fraction of that seven million. Suppose, for instance, that your book is about Spey casting; that means your market is limited to actual and potential Spey casters, and nobody really knows how many there are. Or if you're writing about the virtues of short fly rods, nobody can tell you how many anglers constitute your target market. What can you say in the absence of more precise information? Well, you could point to the number of fly-fishing books already in print, which is evidence of a consistent existing market. Another thing is to point out that fly fishing has become an expensive sport, with fly rods that cost $1,000 or more, waders that cost $600, fly lines at $100 or more, and increasingly popular trips to exotic fishing destinations that cost thousands—all evidence that among the general population of fly fishers are many people with money to spend on books. Those arguments might not carry as much weight as precise, verifiable statistics, but they may help convince a publisher, especially one that has little experience publishing fly-fishing books, that your book is a good investment. And remember, publishers always must consider the bottom line.

Most marketing analyses also include a list of similar books that might compete with yours. Don't dodge the question; spend some time on the internet or in bookstores to see what's selling now, and list potentially competitive books by title, author, publisher, date of publication, number of pages, price, page count and ISBN number. If your survey turns up a number of books similar to yours, you can suggest that's evidence there's already active market interest in your subject, and emphasize any new or original material your book offers that others on your list are lacking.

Some publishers also may ask for a marketing plan, a description of what steps you will take personally to promote your book. If your publisher is one of those, you'll find plenty of information in the next chapter, "Promoting Your Book," to use in preparing such a plan.

Final thoughts: Make sure you're targeting your proposal to the right publisher, one with a history of publishing fly-fishing books. Also be certain your manuscript is of sufficient length for a book—usually at least 50,000 words, though smaller books, if clearly and narrowly focused on a subject of sufficient interest, do have a chance. Some notable books, such as Ed Koch's *Fishing the Midge* (1972) and Sylvester Nemes' *The Soft-Hackled Fly*, were milestones in fishing literature despite their small size and comparative brevity.

Stick to the format mentioned above. If you're submitting by email, the subject line should be clear, precise and brief.

Good luck.

Do You Need an Agent?

If your book proposal fails to win approval from your target publisher, perhaps it's time to seek a professional literary agent. You can use the same proposal, accompanied by a one-page cover letter, to introduce yourself to an agent whose experience might lead to a better result, or one with access to the megapublishers that accept proposals only from agents. But before you do that, there are many pros and cons to consider. Paul, who once had an agent; has a lot to say about both the pros and cons, beginning with debunking a popular misconception: It's an awkward truism that people who have never published tend to imagine that people who do publish make lots of money. To many people, a writer is a writer, and a book is a book; whether your name happens to be Stephen King and you write movie-ready horror novels, or Snavely Berfle and you write obscure fishing books, casual observers often imagine that the money must be great. By virtue of their mysterious creation, books carry the implication of an almost magical degree of authorial prosperity. The reality, that a very tiny percentage of authors are able to make a living at it, is hardly believable. Brace yourself for these sorts of misimpressions among your non-writing friends. Some of them won't even believe you if you try to tell them otherwise.

At first glance, this situation would seem to indicate that all those underpaid, overworked writers surely must need an agent so they can make more money, but that's not how it works. The short answer to the question, Do I need an agent? is almost certainly no, but there's more to say.

Paul's experience was that having a literary agent was great. Simply knowing there's someone who respects your work enough to want to represent it is a serious morale boost, but having such a person check in with you now and then to talk over what you're working on and offer advice and encouragement is priceless. But what makes them so well worth their off-the-top cut of your profits (15 percent, perhaps more) is their savvy in the marketplace. Contracts seem to grow longer and more opaque as the market gets more complicated; 40 years ago who knew there'd be the computer's many digital enticements and complications? It's a great challenge to keep up with all the rapidly evolving aspects of writing and publishing. It's part of an agent's job to do so, and thus to provide informed representation of your interests in light of all that must be known about that side of things.

Having your own personal cheerleader and up-to-date advisor on publishing's byzantine infrastructure is also worth whatever cut the agent takes because agents are so good at getting you more money in the first place. Traditionally, most writers aren't all that hot at bargaining with a publisher; they were (and are) simply too close to the whole thing, and were (and are) too fearful they'll get it wrong and blow the deal completely. By contrast, an agent brings to the negotiation table a comfortable emotional distance. By virtue of his or her background and experience, an agent can instantly translate the obscure language of the fine print, gauge the publisher's interest and rightness for the project, and immediately have a pretty good idea of how much the publisher will—and can afford to—flex on every aspect of the contract, not just the up-front money and royalty rates but serial sales, warranties and indemnities, sharing of foreign reprint sales, paperback and revised edition reprints, legal liabilities, publicity budget, and even—here you dream most extravagantly—film rights. With that expertise and assurance, an agent converses with the publisher's representative in a whole different way: two old hands going through the familiar back-and-forth to reach a mutually acceptable common ground that both parties probably knew was about where they'd end up all along. There are writers who can do that, but not many. Certainly not us.

So considering all those advantages, why might you not need an agent? Well, to beat a dead trout, we'll say this again: Fly-fishing writing is by any financial definition small time. Very few serious professional agents would even look at negotiating with fly-fishing magazines just because the money involved, which probably won't amount to minimum wage for you, will amount to almost nothing for them ("Fifteen percent of $125? Are you kidding me? I couldn't afford to represent that article for less than $500 as my share off the top.")

The same is true for most fly-fishing books. We love our books, but only a few thousand of us are likely to buy most new ones. A manual on how to fish the West Fork of the East Branch has nowhere near the economic *umph* to earn meaningful money, certainly not enough to justify an agent's time. On the usual scale of a fly-fishing writer's income, you will have a harder time, and much less chance of success, negotiating for an agent's representation than you will negotiating with a publisher to publish your book.

As we've also said, there are and always will be a few fishing writers who do well enough to be published by large-circulation magazines and industrial-strength book publishers. Those few writers need and often have agents. Go ahead and dream of reaching that lofty financial eminence as a writer, but in the meantime get to know how to read contracts for yourself, and otherwise learn the business as well as you can. Be realistic, and never make absurd demands on some small-circulation magazine or small publisher. Focus on becoming a skilled and respected writer and, at some time in that process—maybe soon, maybe later, maybe never—you might find yourself with just the right project in hand, one with enough promise to generate some enthusiasm among agents. Then, by all means, have an agent.

Say hi from us.

If, after reading that, you're still convinced you need an agent, the next step is to try finding one, which is somewhat like shopping for a mail-order Russian bride. The latest issue of the *Writer's Market* lists nearly 60 pages of agents and the Agent Query website has even more, plus tips on how to write query letters. It also allows you to search for individual agents or agencies and their specialties, but if you search for "fly fishing literary agents" you'll come up with only a handful of entries; lots of agents list "sports" as a specialty, but that rarely if ever includes fly fishing. Maybe that's good because it certainly narrows your search, but it's also confirmation that fly-fishing authors don't loom large on the horizons of literary agents. But if you want to persist, you can find lots more information about finding and working with agents on the website of Jane Friedman, an experienced and recognized expert on the intricacies of the book-publishing world.

Remember too that agents are paid by commission—the 15 percent or more mentioned earlier—so they are always on the lookout for younger, talented, prolific writers they can count on to be with them for years to come—and the more diversified the writer's portfolio, the better. Some fly-fishing writers have made themselves profitable clients for agents by sometimes writing about subjects other than fly fishing; Paul, for example, has written books about Yellowstone Park and its history, bears, natural history and other subjects in addition to his fly-fishing books. The late Bill Tapply was a great example of a diversified writer; besides his fly-fishing books—*Every Day Was Special* (2014), *Pocket Water* (2001), *A Fly-Fishing Life* (1997) and others—he wrote several mystery series, most notably one featuring Brady Coyne, a fictional Boston attorney whose clients kept getting themselves and Coyne in deep trouble, but never so much there wasn't time for an occasional fishing trip. The list of his publishers is almost as long as his list of titles, and his story drives home the point that if you can demonstrate proficiency writing about more than one subject, you'll make yourself more attractive to an agent and/or publisher.

Keep in mind, though, that you need to be careful in your agent search; as in so many other aspects of our complex modern life, there are predatory scammers out there posing as literary agents and waiting to get access to your personal information and grab some of your hard-earned money, so make certain any agent you contact is an upright member of the Association of American Literary Agents and subscribes to its canon of ethics.

Finding the Right Publisher

Who publishes fly-fishing books these days? Unfortunately, the list isn't very long and it changes rapidly because book publishers, like magazines, come and go frequently. Moreover, book publishers' interests change; tomorrow they may not have any interest in the topics they like today. All the recent consolidations in the publishing industry also make it hard to keep track of the complicated relationships among publishers and their various divisions and imprints. However, we've tried to work through those complications for you, and what follows is an alphabetical list of publishers we have identified that are active currently or occasionally in the field of fly-fishing book publishing, although we can't guarantee we haven't missed one or two. Also, keep in mind the field is a moving target, so our list might already be a little out of date by the time you see it—just like yesterday's fishing reports.

Here's the list:

Amato Media, Milwaukie, Oregon (https://amatobooks.com). Back in the late 1960s a young teacher named Frank Amato yielded to his love for all things fishing and founded Frank Amato Publishing in a Portland suburb, venturing first into magazines, then into books and maps, and began producing a flood of soft- and hard-cover books on all aspects of fishing. These include many types of fishing that most of us would do anything to avoid, but they also include many fly-fishing titles. Amato's start-up enterprise gradually expanded and introduced many new fly-fishing authors to the public, including Dave Hughes, Trey Combs and Randall Kaufmann. Most of his books were of the how-to and where-to variety, and some were criticized for unimaginative design or lack of professional editing, but that didn't seem to hurt their popularity. Amato's books also were promoted in the company's magazines, sometimes cross-promoted in the books themselves.

Amato Media has not posted submission guidelines—at least none that we have been able to find after diligent searching—so our best advice to prospective authors is to study the titles listed on the company's website and perhaps some of the books themselves to get an idea of the types of fly-fishing books it publishes, then prepare and submit a book proposal.

Chattahoochee Media Group, Roswell, Georgia (flykits.net). This publisher (which also markets fly-tying kits) has mostly published regional paperback fly-fishing guide books written by its founder and editorial director, Steve Hudson. Examples include *Chattahoochee Trout: The Definitive Guide to Chattahoochee Trout Fishing* (2017) and *Fly Fishing Georgia's Toccoa River* (2015). Their website indicates they would consider submissions by other writers, but the submission instructions are a bit confusing: "We appreciate a short cover letter with a concise bio that could be used should we publish the work," but "please do not mail or email submissions." Maybe you're supposed to stop by and drop off your proposal. Good luck.

Hancock House, Surrey, British Columbia (submissions@hancockhouse.com). Most of the fishing titles from this Canadian publisher are getting old, but a new one—*A Man and His River*, by D.C. Reid (2022), might be a sign they're ready for some other new material. Reid's book is "a love story" about the Nitinat River on Vancouver Island and his fishing experiences in some of the many other rivers on that island. Hancock House also posts unusually detailed and helpful submission guidelines on its website, starting with this statement:

"In accordance with our environmental objectives, we prefer not to waste paper; therefore, initial queries should be made via email." They don't publish fiction. Book proposals should include *three* sample chapters, an annotated table of contents, a 250- to 300-word synopsis in Microsoft Word or PDF, the total word count and number and type of images in the book, a short author biography and qualifications, a list of similar books with an explanation of why yours is "different or more relevant," and a description of your book's target audience.

Take note that the website also advises that Hancock House looks more favorably on "projects where the author is prepared to take on a leading role in the promotion and marketing of their book." It also poses this interesting question: "Why have you sought to contact us for publishing your work as opposed to another publishing house?" Hint: Don't tell them it was because you couldn't find another publisher willing to publish your book.

Book proposals submitted to Hancock House "normally receive a response within one to two weeks," though it's "not unusual for the process to take longer." If they're interested in your manuscript, they will request a complete copy to review—and "we may also make suggestions as to how your manuscript can be changed to make it more publishable."

Headwater Books, Boiling Springs, Pennsylvania (headwaterbooks@gmail.com). Their website advises that Headwater Books "is a publishing company specializing in fly-fishing and fly-tying books written by experts. Our authors are leaders in their fields and passionate about sharing their knowledge and expertise." . . . Through a partnership with Stackpole Books (see below) Headwater Books "provides the distribution and marketing of a large company with the service of a small one." Maybe just what you've been looking for, right?

The editor at Headwater is Jay Nichols, a respected editor and writer who cut his editing teeth in previous service at *Fly Fisherman* magazine. He also wears a second hat as fly-fishing editor for Stackpole Books. Headwater prefers queries the old-fashioned way, on paper, which should be mailed to P.O. Box 202, Boiling Springs, PA 17007. They should include a brief description of your book idea, including the working title and estimated length, "two or three" sample chapters, a detailed outline and table of contents, a "brief overview of the market for your book and its competition," a brief biography with emphasis on your publishing experience and qualifications to write the book, complete contact information, and perhaps a thumb drive with samples of your photos or other artwork. Jay will take it from there.

Some of Headwater's recent titles (which all seem to have long subtitles) include *Game Changer: Tying Flies that Look and Swim Like the Real Thing*, by Blane Chocklett (2020); *Keystone Fly Fishing, The Ultimate Guide to Pennsylvania's Best Water*, by Henry Ramsay, Mike Heck and Len Lichvar (2017), *The Bug Book, A Fly Fisher's Guide to Trout Stream Insects*, by Paul Weamer (2016), and

Selectivity, The Theory and Method of Fly Fishing for Fussy Trout, Salmon and Steelhead, by Matt Supinski (2014).

Keokee Books, Sandpoint, Idaho (books@keokee.com). This small regional press, founded in 1990, publishes books on a variety of topics including fly fishing. It also provides services to authors engaged in self-publishing. Recent fly-fishing titles include *Chasing Ghost Trout*, by Dennis Dauble (2021), chronicling the author's pursuit of wild trout in the streams of Oregon's Blue Mountains. Notable past titles include Mallory Burton's *Reading the Water: Stories and Essays of Fly Fishing and Life* (1995), and Paul Quinnett's *Pavlov's Trout* (1994). Keokee publishes only two or three titles a year and its website advises that "manuscripts are accepted by invitation only." It doesn't say how to get on the invitation list.

Lyons Press, Guilford, Connecticut (imprint of Globe Pequot, Rowman & Littlefield) (editorial@lyonspress.com). Nick Lyons, an English professor at Hunter College in New York, took over the reins of Crown Publishing's *Sportsman's Classic* series in 1970 and quickly turned it into the greatest powerhouse of fly-fishing book publishing the sport has ever witnessed, producing new titles and reprints of older fishing classics in unprecedented numbers. One of the earliest, in 1971, was *Selective Trout*, by Carl Richards and Doug Swisher, which became the best-selling fly-fishing book of its time. That same year brought forth *Fishless Days, Angling Nights*, by Sparse Grey Hackle (pseudonym of Alfred W. Miller), which became another classic. Other titles included *Fisherman's Bounty* (1970), a huge anthology edited by Lyons himself, plus reprints of Jennings' *A Book of Trout Flies*, all four "Seasons" books by Roderick Haig-Brown, and Vincent Marinaro's *A Modern Dry-Fly Code.*

In 1977 Lyons left Crown and founded his own publishing company, Nick Lyons Books, as a subsidiary of Benn Brothers Limited, a British publisher. Nick Lyons Books published a variety of books, but fly-fishing was always a high priority for Lyons, himself an avid angler. His long list of titles, some issued in conjunction with Winchester Press (more about them later), included Gary LaFontaine's *Caddisflies* (1981), Jacqueline Wakeford's *Flytying Techniques* (1981)—one of few fly-fishing books written by women to be published at that time—Dave Whitlock's *Guide to Aquatic Trout Foods* (1982), Austin Francis' *Catskill Rivers* (1983) and W. D. Wetherell's *Vermont River* (1984). Lyons bought out the subsidiary rights from Benn in 1984, then published a revised edition of Lefty Kreh's *Fly Fishing in Salt Water*, Richard Talleur's *The Fly Tyer's Primer*, and Lee Wulff's *Trout on a Fly*, all in 1986; Joan Wulff's *Flycasting Techniques* (1987), and many more. In 1989 he took on a partner, Peter Burford, and changed the name of his firm to Lyons & Burford. Together they published another fine list of fly-fishing books, including *Steelhead Fly Fishing* by Trey Combs (1991), *Notes from the San Juan* by Steven J. Meyers (1991), *The Battenkill* by John Merwin (1993), and *The Habit of Rivers* by Ted Leeson (1994), plus a number of reprints, including Haig-Brown's *A River Never Sleeps* and *Return to the River* and John McDonald's *The Origins of Angling*. Burford left the partnership in 1997 and the company's name was shortened to the Lyons Press, with Nick Lyons remaining at the helm. Subsequent titles include several by women writers—Joan Wulff's *Fly Casting Accuracy* (1997), Jessica Maxwell's *I Don't Know Why I Swallowed the Fly* (1997), Mallory Burton's *Green River Virgins and Other Passionate Anglers* (2000) and Joan Wulff's *New Fly Casting Techniques* (2012).

Now the Lyons Press is an imprint of Globe Pequot, a division of rapidly metastasizing Rowman & Littlefield. Lyons Press remains a worthwhile possibility for aspiring or experienced fly-fishing authors; it also remains friendly to women writers. Recent titles include *A Sportsman's Life, How I Built Orvis by Mixing Business and Sport* (2022, a new edition of a book originally published in 1999 by The Atlantic Monthly Press), by the late Leigh Perkins. The Lyons Press also publishes guidebooks for Orvis and L.L. Bean; some are staff-written but others might offer opportunities for freelancers.

Submission requirements are generic to a long list of imprints, but those for the Lyons Press should be addressed to editorial@lyonspress.com. Email submissions are preferred, but "anything received in the mail will be recycled after it is considered." Proposals should include a description of the book, table of contents, sample chapter and sample photos or illustrations, a summary of the author's publishing experience and a description of his or her "online presence"—an explanation of how you will help promote the book. They should also include a market analysis including answers to the following questions: "Who will buy this book?" and "Why publish this book now?" The analysis also should include a list of current competitive titles.

No Nonsense fly fishing guidebooks, Tucson, Arizona (nononsenseguides.com). Guidebooks have become so popular they now practically constitute a subgenre of fly-fishing books; they aren't usually a source of fine literature, but they do offer many opportunities for writers, and No Nonsense is perhaps the leading publisher of such books. Its website has no submission guidelines but does offer the rather general statement that "No Nonsense fly-fishing guidebooks give you a quick, clear understanding of the essential information needed to fly fish a region's most outstanding waters. Our authors are experienced and qualified local fly fishers. Maps are fully researched using a variety of sources to ensure that you have the latest and most accurate information." So if you're an experienced and qualified local fly fisher, maybe this publisher is for you.

No Nonsense titles include *Guide to Fly Fishing in the Western States*, by Bob Zellere; *Fly Fishing the Mid-Atlantic*, by Beau Beasley; *Fly Fishing Central California*, by Brian Milne; *Fly Fishing Georgia*, by David Cannon, and a host of other state and regional guidebooks, probably including one or more for the area where you live. They also apparently publish occasional "how-to" guides, such as *Become a Thinking Fly Tier*, by Jim Cramer. To get in touch, email NancyFisher@nononsenseguides.com.

Patagonia Books, Ventura, California, Karla Olson, publisher (patagonia.com). You might be surprised to find Patagonia on a list of publishers, but in addition to all its gear and clothing manufacturing, environmental activities and philanthropy, Yvon Chouinard's fascinating company (recently donated to the nonprofit Holdfast Collective to fight climate change) also produces occasional books, including some about fly fishing. The latest, as of this writing, is *Headwaters: The Adventures, Obsessions and Evolution of a Fly Fisherman*, by Dylan Tomine, a "Patagonia ambassador," whatever that is (2022). Another—although not strictly a fly-fishing book—was *Salmon: A Fish, the Earth and the History of Their Common Fate, by* Mark Kurlansky (2020). Patagonia does not post submission requirements for its books, but if you search the internet for "Hope and Action: The Mission of Patagonia Books" you'll find a long, interesting and helpful interview with Karla Olson, publisher, which explains what they are interested in publishing, how they acquire books and their future publishing directions. As ever, it's also a good idea to check out some of their other fly-fishing books.

Rizzoli International Publications, New York, New York (submissions@rizzoliusa.com). Rizzoli, part of the Mondadori Group, Italy's largest publisher, has offices and a bookstore in New York City and has been publishing a diverse array of books since 1974. Its fly-fishing titles include a beautiful coffee-table book titled *Trout*, by Tom Rosenbauer (2022)—and yes, that makes at least three different books with the same title—and *Great Fishing Lodges of North America: Fly Fishing's Finest Destinations*, by Paul Fersen (2010)

According to its website, Rizzoli has no set format for book proposals but "it would be helpful to include an outline, sample chapters, sample designs, and/or sample art." Rizzoli prefers digital submissions, as long as they don't exceed five megabytes. Be certain to include your email address and other contact information. It also accepts submissions by regular mail, but you should provide a self-addressed envelope and postage if you want your material returned. Mailed submissions should be addressed to Rizzoli International Publications, Attn.: Editorial Submissions, 300 Park Avenue South, Fourth Floor, New York, NY 10010. Allow up to 12 weeks for a response, but if you provide an email address you may receive a more prompt reply. "Submissions will be returned within the twelve weeks unless you are directly notified by an editor."

Skyhorse Publishers, New York, New York (skyhorse publishing.com). Nick Lyons' son, Tony Lyons, former president of the Lyons Press, established his own business, Skyhorse Publishing, in September, 2006. Skyhorse soon became one of the fastest growing independent book publishers in the United States, accumulating various important imprints and publishing both fiction and nonfiction in a wide variety of topics, but always prominently including fly fishing. Some of its fishing titles include *The Complete Book of Fly Tying*, by Eric Leiser, C. Boyd Pfeiffer and Jack Gartside (2014), *The Fly Fisherman's Guide to the Meaning of Life*, by Peter Kaminsky (2008), Lefty Kreh's autobiography, *My Life Was This Big* (2014), *A Fly Fisher's World*, by Nick and Mari Lyons (2013), *Trout Fishing in the Catskills*, by Ed Van Put (2014), *Angling Days: A Fly Fisher's Journals*, by Robert DeMott (2016), and Maxine Atherton's *The Fly Fisher and the River* (2016).

Skyhorse posts its submission guidelines on its website. They include a brief query letter, a one- or two-page synopsis of your proposed book, an annotated chapter outline, a market analysis, one or two sample chapters, and your biography, including a list of all previous published work. Email submissions to submissions@ skyhorsepublishing.com (hard copy submissions are not considered). The subject line of your email should say "outdoor & sports" (one of a number of subject category lines listed on the website, along with the warning that Skyhorse will not review any proposals with subject lines lacking one of the categories on the list.) "If we are interested we'll get back to you in four to six weeks."

Stackpole Books, Mechanicsburg, Pennsylvania (stackpolebooks.com). Venerable Stackpole Books is another top publisher of fly-fishing books in the United States. It was founded in Harrisburg, Pennsylvania in 1930 by brothers Edward James and Albert Stackpole and began publishing books on military history, including several written by Edward James Stackpole himself (he was a general in World War I). Later the company began publishing books in many other categories, both nonfiction and fiction, including fly-fishing books. Notable titles included Alvin R. ("Bus") Grove's *The Lure and Lore of Trout Fishing* (1951), Sid Gordon's *How to Fish from Top to Bottom* (1955), Charles M. Wetzel's *Trout Flies, Naturals and Imitations* (1955), Lee Wulff's *Lee Wulff on Flies* (1980), and Joan Wulff's *Fly Fishing: Expert Advice from a Woman's Perspective* (1991). Under the guidance of highly regarded editor Judith Schnell, the company continued to build its distinguished list of fly-fishing books, aided by its partnership with Headwater Books (see above), and Jay Nichols is now Stackpole's fly-fishing editor. In 2015 Stackpole was acquired by Rowman & Littlefield Publishing, and now, as an imprint of Globe Pequot Books with Judith Schnell as publisher, it continues publishing fly-fishing books under its own label.

Stackpole receives book proposals either by mail or email. Email submissions should be sent to editorial@stackpolebooks.com. All submissions should include your mailing or email address. The proposal should include a description of the book, table of contents, sample chapter, samples of photographs or illustrations, a recital of your qualifications to write the book, your previous publishing experience, and your "online presence" (how you will help promote the book). It also should include a market analysis—who will buy the book, why it should be published now, and a list of competitive titles. Allow three months or more for responses.

The Whitefish Press, Cincinnati, Ohio (whitefishpress@yahoo.com). According to its website, the Whitefish Press "is a radical departure in publishing. Not only do we produce high quality books at affordable prices, our royalty-share program offers authors up to 50 percent of the royalties for their original works.... We are not a print-on-demand company, but a full-scale publisher working with one of the premier book binderies in the nation. Additionally, we offer authors unprecedented control over design, layout, promotion, etc., as well as professional editing at all stages of a project. And to top it all off, our staff includes some of the most knowledgeable people in the field of fishing history in the world. We handle the editing, layout, publishing, and distribution of your work. You put up NO money and get up to 50 percent of the royalty. Try to find a better deal than that!" We'll wait a moment for the excitement to subside before we continue.

The Whitefish Press, an imprint of Micropress America L.L.C., is the brainchild of Dr. Todd E.A. Larson, a history professor at Xavier University, who "writes on the history of fishing in America." Founded in 2006, it has worked to grow the business "to the point where it will be a unique voice in the publishing world for those interested in preserving our outdoor heritage." Not all its books are about fly fishing, but a number of them are. They include *Casting About: The Complete Castabout Columns on Fly Rods and Fly-Fishing History*, by Jerry Girard (2022), which includes 50 columns gathered from magazines and other sources by the Catskill Fly Fishing Center and Museum; *The Lew Stoner-Ted Trueblood Letters* (2021), a limited edition work featuring letters between Trueblood, a famous outdoor writer, and Stoner, an equally famous and innovative bamboo rod builder; *Howells: The Bamboo Fly Rods & Fly-Fishing Legacy of Gary H. Howells*, by Joseph H. Beelart, Jr. (2020), a biography of one of American's greatest rod builders; and other similarly intriguing books.

Whitefish also has published several works by women writers, including two by Erin Block, *The View from Coal Creek: Reflections on Fly Rods, Canyons and Bamboo* (2012) and *By a Thread: A Retrospective on Women and Fly Tying* (2016), and *The Origins of American Angling: Essays on the History of American Fishing and Tackle*, by Mary Kefover Kelly (2007).

Whitefish's website invites those who "have a manuscript, outline, or an idea you think we might be interested in" to contact them. Just make sure the subject line of your email says ATTN: Marc Hanger, Acquisitions editor.

Wild River Press, Bothell, Washington (wildriverpress.com). Tom Pero has been involved in fly-fishing magazine and book publishing most of his life, but his current venture is the Wild River Press. On his website he explains "when I started Wild River Press, my aim was simple but ambitious: Produce the finest books about fly fishing that I could. I didn't want to produce a lot of books, rather a few good ones. And now, several years later, including a couple of national book awards, I think it's fair to say we have done that." That's fair to say, all right; Wild River has produced some large and often spectacular books, including limited editions. Representative titles include *Fifty Women Who Fish* by Steve Kantner (2020), *A Passion for Permit* (two volumes) by Jonathan Olch (2016), and *Atlantic Salmon Magic* by Topher Browne (2011).

Wild River does not post submission guidelines, but this quote from Pero's website will give you a pretty good idea of the type of material they publish: "Our books are authentic. Our books have compelling narrative. Our books also look good. They are well-written and crisply edited. They are beautifully designed and handsomely crafted. My philosophy is: Exceptional books should be both literate and visually pleasing. There are too many forgettable fishing books in print." Take it from there.

Wilderness Adventures Press, Belgrade, Montana (wildadvpress.com). Founded in 1992, this mom-and-pop publisher has been publishing state and regional fishing guidebooks ever since. Representative titles include *Flyfishers Guide to Arizona*, by Will Jordan, *Field Guide to Fishing Knots*, by Darren Brown, *Washington's Best Waters* (a book of maps), and many similar books. Submission guidelines are not posted, but the company's website has this to say: "We created a series of state and regional guidebooks that would give outdoor enthusiasts complete and up-to-date information that would make their fly fishing, hunting, or birding adventures a success and a memorable trip. We seek out authors for our books who have a passion for their sport and outstanding knowledge of the area that they are writing about." All their books have detailed maps and they also publish river maps with GPS coordinates in full-color map books.

Wind Knot Publishing, Burlington, Vermont (windknot publishing.com). This small publisher publishes fishing guidebooks focused mainly on the state of Vermont, mostly written by Peter Shea, who just happens to own the company. Titles include *Vermont's Trophy Trout Waters*, by Shea; *Vermont Trout Ponds*, by Shea, and *Long Trail Trout*, by—you guessed it—Peter Shea. No submission guidelines are posted, but if you to know something about fly fishing in Vermont—or if your last name happens to be Shea—well, there you go.

Missing in Action: Two once very prominent publishers of fly-fishing books are unfortunately no longer with us, but their contributions to the sport remain worth mentioning. Winchester Press, originally a division of Winchester firearms, was founded about 1968 in New York and had several different ownerships in several different locations during its approximately 20 years of existence, but during that time it published a remarkable series of fly-fishing books, some in conjunction with the Lyons Press. Many of those books remain important, including the first edition of Ernest Schwiebert's *Nymphs*, Arnold Gingrich's massive *The Fishing in Print* (both in 1974), and Martin Keane's *Classic Rods and Rodmakers* (1976). Pruett Publishing, of Boulder, Colorado, was founded in 1954 and came late to fly fishing, but one of its efforts was a blockbuster, *Trout Bum* (1986), the first book by John Gierach, starting him on a series that has taken him all the way into the sanctified precincts of publishing giant Simon & Schuster. Other significant titles included *A Trout's Best Friend*, the autobiography of Bud Lilly (assisted by a familiar name, Paul Schullery) (1985), and *Rivers of*

Memory, the last book of the late Harry Middleton (1993) published during his life. Pruett was sold in 2012.

University Presses

There are a great many university presses in North America, but traditionally they have rarely descended from their academic preoccupations to publish anything about a subject as unacademic and fun as fly fishing. But in 1976, the University of Chicago Press did, and it almost turned the world of fly fishing on its ear. The book was *A River Runs Through It*, by Norman Maclean, which became the inspiration for the later movie starring Brad Pitt, which triggered a stampede of new fly-fishing converts into tackle shops and crowded rivers, as well as a bit of a boom in the Montana real-estate business. However, prospective authors who think of selling their book to a university press should keep in mind that Maclean was a distinguished member of the University of Chicago faculty, and it's certainly a lot easier to get a university press to publish your work if you're on the faculty. Being an alum of the university also might help.

That said, following are some university presses (or their imprints) that have published books about or related to fly fishing. Some reflect what appears to be a promising trend to link the sport with public policy, environmental and other issues that appeal to university publishers.

University Press of Colorado, Boulder, CO, published Gordon Wickstrom's *Notes from an Old Fly Book* in 2001.

University of Georgia Press, Athens, GA, published *Searching for Home Waters: A Brook Trout Pilgrimage*, by Michael K. Steinberg in 2023.

University of Nebraska Press, Lincoln, NE. Under its Bison Books trade imprint, the University of Nebraska Press published Quinn Grover's *Wilderness of Hope: Fly Fishing and Public Lands in the American West* (2019), which describes how public lands and waters largely made possible his fly-fishing career.

University of New Mexico Press, Albuquerque, NM. As part of its Coyote Books series, the University of New Mexico Press has published several guidebooks, including *Fishing in New Mexico*, by Ti Piper (1989); *Fly Fishing in Southern New Mexico*, by Rex Johnson and Ronald Smorynski (1998); and *Fly Fishing in Northern New Mexico*, edited by Martin Craig (2002). It also published Gordon Wickstrom's *Late in an Angler's Life* (2004), and new editions of several of Paul's books, including *Royal Coachman*, originally published by Simon & Schuster in 1999.

Ohio University Press, Athens, OH, published *Haunted by Waters: Fly Fishing in North American Literature*, by Mark Browning (1998), and under its Swallow Press trade imprint produced *Yellow Stonefly*, by Tim Poland (2018), a rare novel from a university publisher.

University of Virginia Press, Charlottesville, VA. Though this press seems not to have published fishing books recently, it was quite active in outdoors publishing for a while especially with outdoors-oriented guide books for its region but also with fly-fishing books specifically, such as Stanley Ulanski's *The Science of Fly Fishing* (2003) and James Barilla's *West with the Rise* (2006).

University of Washington Press, Seattle, WA, published *Trout Culture: How Fly Fishing Forever Changed the Rocky Mountain West*, by Jen Browne (2015), which describes how "fish culturalists and anglers" manipulated the natural Rocky Mountain environment by introducing new fish species and stocked billions of hatchery trout in western waters.

University of Wisconsin Press, Madison WI. This press has a history of publishing fishing books, including *Exploring Wisconsin Trout Streams: The Angler's Guide*, by Steve Born, Bill Songnozi and Jeff Mayers (1997); *Catching Big Fish on Light Fly Tackle*, by Jeff Mayers and Tom Wendleburg (2001); and *Sunlit Riffles and Shadowed Runs, Stories of Fly Fishing in America*, by Kent Cowgill (2012), a collection of 16 tales.

Yale University Press, New Haven, CT. Yale University, whose library holds one of the world's premier angling-book collections, would at first glance seem a likely publisher for the occasional fishing book, but their production of such books has been small. However, they did publish Anders Halverson's *An Entirely Synthetic Fish* (2010), an engaging account of how the rainbow trout has become the most widely stocked and domesticated trout in existence.

Those are only a few of the university presses; there are lots of others. The problem—if that's what it should be called—with university presses as angling book publishers remains in good part one of academic snootiness. Though fishing has made gratifying strides in recent decades as far as being perceived as a subject fit for serious mainstream scholarly study, even the few university presses that have taken the plunge into fishing literature seem to be feeling their way, and might be comfortable only with such books as relate directly to their university's own community or special circumstances.

Megapublishers

Every once in a while one of the big conglomerate publishing houses will surprisingly produce a fly-fishing book, but if you want one of those megapublishers to publish yours you'll need an agent to make the pitch, because nowadays few if any of the big publishing giants will condescend to deal with mere mortal writers. That said, what follows is a partial list of megapublishers that published fly-fishing books in the past, mostly well in the past; perhaps your agent can convince one of them it's about time for another.

Alfred A. Knopf, now part of the **Penguin Group**, was for much of the 20th century among the most important publishers of distinguished, authoritative outdoor books, under its Borzoi Books imprint. A light sampling of that great run of fine books would include Arnold Gingrich's *The Well-Tempered Angler* (1966), John McDonald's *Quill Gordon* (1972), Joe Bates' *Streamers and Bucktails* (1980), Datus Proper's *What the Trout Said* (1982), Thomas

McGuane's *The Longest Silence* (2000) and Henry Hughes' *The Art of Angling: Poems about Fishing* 2011).

Macmillan Publishers, now part of the German-owned **Holtzbrinck Publishing Group,** in 1955 published the first edition of Ernest Schwiebert's *Matching the Hatch*, another landmark 20th-century book.

Pegasus Books in 2022 published *Illuminated by Water: Fly Fishing and the Allure of the Natural World*, by Mallachy Tallack.

Rowman & Littlefield in 2006 published William Rutter's *Basic Essentials: Fly Fishing.*

William Morrow & Co., now an imprint of **HarperCollins,** in 1946 published the first edition of Roderick Haig-Brown's *A River Never Sleeps* and many other books by him.

G.P. Putnam Sons, now an imprint of giant **Penguin Group**, published Vince Marinaro's *A Modern Dry Fly Code* (1950), one of the most influential fly-fishing books of the 20th century.

Charles Scribner's Sons, now part of **Simon & Schuster**, in 1947 published John McDonald's *The Compleat Fly Fisherman: The Notes and Letters of Theodore Gordon.*

St. Martin's Press, now part of **Macmillan** and **Holtzbrinck Publishing**, published Robert Traver's *Trout Madness* in 1960.

And there are others.

But if all else fails, see chapter 12 on self-publishing.

Coping with Your Contract

When you do finally hook up with a publisher and receive a contract, it's definitely a cause for celebration. Break out the champagne and wrap yourself in the great feeling of satisfaction and happiness that your long writing journey has finally reached a happy ending.

Or has it? When you take your first good look at the contract you might be overwhelmed. It's so long! It's so complicated and full of legal gobbledygook! And no wonder, because you're looking at the work of a whole nest of busy lawyers who toted up many billable hours. It can be very intimidating, but Paul and Steve have seen enough contracts to make at least some sense of the elements most have in common. This is by no means an in-depth analysis though, just a few brief tips and suggestions.

First, a couple of things to keep in mind while you peruse all that legalese. As we've said before, book contracts are always weighted heavily in favor of publishers, to the point that your initial inclination might be that you don't want to sign such a document. Remember, however, the publisher has a lot more at risk here than you do; it's the publisher who's putting up the money to pay you an advance and royalties, and the publisher who's footing the bill for editing, design, printing, binding and publicizing your book, all in the hope of generating a profit. The contract is a legal agreement that spells out the responsibilities and obligations of both parties, including your warranty that the book (usually referred to as "the work" in contracts) is your own original work and does not infringe on any other copyrights or include any libelous material. It also sets forth remedies available to each party in the event that either violates terms of the agreement, and these also usually weigh heavily in favor of the publisher, so you want to be sure you abide by them. It also establishes the amount and timing of the advance you will receive and sets a scale of royalties for each type of sales of the book (such as discounted or promotional sales, etc.), the sale of subsidiary rights (such as film, dramatization, anthology, audiobook, paper or hardback reprints, etc.), and establishes procedures for accounting, reporting and paying royalties, and lots more or less essential stuff.

If all this makes you think you need a lawyer just to interpret the contract, you can certainly hire one, but keep in mind that yours is probably a niche book that won't earn a lot of money, possibly even less than an attorney would charge to review the contract. Membership in the Authors Guild entitles you to a review of the contract by a Guild attorney, but you have to be a published author to qualify for membership, so that avenue might not be open to a first-time author. So likely you will need to handle it by yourself, and despite the length and complexity of the contract language, you can probably do that yourself if you read it slowly and carefully.

Even so, you might see some things you don't like. What to do then? Probably nothing, unless it's a really serious problem, a real showstopper. Remember, if this is your first book, you have very little leverage and don't want to make the publisher angry; it could result in withdrawal of the contract. So it's best not to quibble. That doesn't mean you shouldn't ask questions, however, and if you see anything in the contract that you don't understand or really don't like, you definitely should ask questions or seek clarifications. That also will let the publisher know you're taking this business seriously; just make sure you ask your questions in a civil manner.

Many first-time authors don't have a clear understanding of how advances and royalties work. Advances traditionally were paid to authors who sold publishers the idea for books that were yet to be written, the idea being that a little up-front money would keep the author from starving while he or she finished writing the book. Nowadays it's more likely the publisher will be buying a book that's already written, but the custom is still to pay the author an advance as temporary compensation for work already done. Typically the advance will be half the total amount the publisher is willing to pay for the book, with the first half payable on acceptance of the manuscript and the second upon publication of the book.

Since fly-fishing books usually don't make much money for either the publisher or author, the advance may not be very large, but the publisher will still expect you to pay it back. For illustrative purposes let's assume your publisher has offered an advance of $5,000, half payable on acceptance of the book and half on publication. But after the book is published and the royalties start flowing in, all will go directly to the publisher until the $5,000 advance is fully repaid; only then will you begin receiving royalty payments. That means it will take months or even years before you receive your first royalty payments, so you'll have to make do with the advance until then.

Before you sign the contract, you also should try to find out what your publisher intends to do about promoting your book, and what's expected of *you*. Most sizeable publishers have publicity staffs that traditionally arranged and scheduled bookstore signings, readings and other personal appearances for authors, but some publishers now try to avoid that by asking you to do it yourself. If your prospective publisher is one of those, that can be an issue of sufficient importance you should try to convince the publisher otherwise—perhaps even at the cost of refusing to sign the contract. Why? The problem is that when authors are asked to arrange their own events, bookstores usually perceive that as evidence the author has been abandoned by the publisher and their response almost inevitably will be, so sorry, but our events calendar is already full for the next two years, which means both you and the publisher lose the chance to promote the book. The publisher should already know this, but you may have to try to explain it and attempt to negotiate a better arrangement—one that will serve both your interests.

Don't let these cautions interfere with your celebration, though. Have more champagne! And maybe some oysters, too. You've earned it!

The torrent of new fishing books is continuous, at last count exceeded only by those devoted to sex.

—Robert Traver (John Voelker), *Trout Magic*

CHAPTER ELEVEN

Promoting Your Book

Books don't sell themselves. As soon as they are off the press they disappear into a complex system that remains a mystery to most people, including authors, until they finally reappear on bookstore shelves and websites. As this complicated system has evolved, authors have become increasingly important, so just when you thought your work was finished and you could relax a bit, your job is only beginning: Now you have to become a publicist and promote your book.

Why? The reason is simple: People won't buy your book, or any other, if they've never heard of it. And because—as we keep emphasizing, if only to make sure you remember—fly-fishing books are aimed at so-called niche markets. Even the largest publishers aren't going to invest heavily to publicize them; instead, the publisher will leave that mostly up to you.

That means you must remember that while modesty is always a good quality, it's sometimes necessary to suppress it temporarily—especially if you want to sell your book. That's because your task is to make sure as many people as possible can learn about your book in as many different ways as possible. Never doubt that this work, as hard as it may be and as distant as it is from what you might once have imagined being a writer was all about, is still useful. It's not just a matter of selling your book; it's a matter of building a readership that will also be looking for your next book, a readership that will permanently care about what you have to say, and will want to know your views for years to come. It's a matter of getting known among all the other parts of the fly-fishing world that aren't directly involved in publishing but still have a lot to do with the fate of a writer's work.

A big-time general-audience nonfiction writer once told Paul that after spending several years writing and traveling the world while researching his most recent book, when it was written he stopped writing for an entire year to promote it; that's a year of not writing at all. Admittedly, that's how it may work in the best-seller big leagues while fly-fishing publishing is more like AA ball, but the same principle applies. Selling your book takes all the work you can give it.

You might not have realized it at the time, but you already may have become involved in the process of promoting your book even before it was published, by providing information for several lists maintained by your publisher.

Media Lists, Blurbists, and eAVCs

Most large publishers keep lists of print and online magazines, newspapers, websites, bloggers, podcasts, television and radio contacts, or other information providers. They use these lists to send press releases, promotional materials and/or review copies of books to selected outlets they believe are most likely to be interested in a particular book. For example, the media list for a new fly-fishing book almost certainly would not include *Sky & Telescope* magazine, or *American Patchwork and Quilting*; instead, it would have an up-to-date list of print and online fly-fishing magazines and other media outlets catering to fly fishers. But it's unlikely a publisher's media lists also would include your hometown newspaper, radio station or TV station, so that's the type of information a publisher's publicist is looking for when he or she asks an author to add to their media list. Chances are your publisher already has done that; if not, it's high time.

Information sought by the publicist may include more than just the names and addresses of hometown information providers; it also might include your university or college alumni magazines; your fraternity's or sorority's magazines; magazines published by professional associations, or magazines that previously published your work. You might even want to add the monthly newsletter of your local fly-fishing club; the publisher isn't likely to send a review copy of your new book to the newsletter editor, but it might send a press release.

Remember, the goal here is to make sure as many people as possible have an opportunity to learn about your new book, and whatever the publisher can do to reach people means a little less you'll have to do yourself, and whatever you can do yourself means less work and cost to the publisher.

Your publisher also may have asked you to provide a list of "blurbists." If that's already happened, then you know what a "blurbist" is (or maybe you did anyway). If not, then it's time we defined it: A "blurbist" is someone who provides one of those one-paragraph testimonials, or "blurbs," on the back cover of a book that tells prospective readers what a swell book it is (blurbs are known in more formal publishing circles as "encomiums"). The best "blurbists" are those with significant name recognition—well-known fellow authors, or, in your case, well-known fellow fly fishers, or, preferably, both. If your publicist is on the ball, she should have asked you well in advance of your book's publication to recommend a list of people you think might give your book some good blurbs.

Who should you choose? It might help if you choose people you know personally, but that's not always possible or necessary. Good blurbists also can be found by scanning the back covers of other fly-fishing books to see who's written generous blurbs for their authors, then add their names to your list. Or you can just choose someone you admire, either as a writer or a fly fisher. Some people feel flattered when asked to write a blurb, and if you're ever asked, don't refuse; you might need to have the favor returned sometime in the future.

Don't be shy about asking someone to write a blurb for you, either. The best writers know how important it is that they return all the favors they've been given on the way to their current prominence. Maybe it won't work out; maybe they really just won't have the time right then (and keep in mind that publishers often don't give blurbists all that much time). But they'll take the request seriously. In rare cases, the author or the publisher may even have connections beyond the normal realm of fly-fishing writers. When Paul helped Bud Lilly write his *Bud Lilly's Guide to Fly Fishing the New West* (2000), Bud reached into his list of celebrity clientele and got a front-cover blurb from world-famous broadcaster (and angler) Tom Brokaw. There are a lot of very famous people out there who love to fly fish. Think big.

Once you've submitted your list, your publisher might offer to contact the proposed blurbists on your behalf and ask if they would be willing to receive and read an eAVRc of the book and write a blurb for it; potential blurbists are more likely to be responsive if asked by the publisher instead of the author. But if the publisher won't do that, you may have to contact the blurbist yourself and arrange for the publisher to send an eAVRc.

OK, so what's an eAVRc? The term is an abbreviation for "electronic advance reading copy." These used to be known just as advance reading copies, paperbound copies of a book's uncorrected page proofs (which usually looked just like a paperback of the book, sometimes even with a low-quality reproduction of the eventual cover) sent to blurbists or reviewers so they could read the book and form an opinion for a review or provide a succinct quote for a blurb. Since the digital revolution has made it easier, faster and less expensive for publishers to send electronic copies of uncorrected proofs, they are now called eAVRcs.

Once an eAVRc of your book has been sent to the people named on your list of blurbists, they are expected to read it, write a blurb and send it to the publisher so the blurb—assuming it's positive—can be placed on the back cover of your book's dust jacket, or maybe on the cover itself. Blurbs also may be used in other promotional materials, including press releases, the publisher's catalog, and information sheets sent out with the review copies.

People asked to prepare blurbs are usually under no obligation to respond, however, so if someone on your list doesn't respond it could mean he or she might have been out fishing, or didn't think your book was worth a blurb (perish the thought), or perhaps submitted a blurb your publicist did not think good enough to display on the back of your book.

Not everyone is a fan of blurbs, either. Paul has worked with editors who considered them completely unnecessary in promoting a book, or as serving only as ego-boosters for the author. We don't see it that way, but it's worth keeping in mind; very few parts of book promotion are sure things.

Remember, all this is going on *before* your book is published. The eAVRcs also are going out to potential reviewers, who, your publisher hopes, will write favorable reviews to be published after your book is in print. Depending on the reviewing medium involved—magazine, newspaper, podcast or whatever—that could be either a short time or a long one. Sometimes you may have to wait a year or more before a printed review appears.

Book Reviews

Your publisher's media list will certainly include magazines that publish book reviews, and reviews have always been one of the very best ways to gain publicity for new titles. Unfortunately, few fly-fishing magazines still publish as many as they once did, and some of those so-called reviews are nothing more than verbatim copies of promotional material printed on the book's dust jacket. There are exceptions, however; *Fly Fisherman* magazine still publishes reviews, though not as often as it once did, and most lack the penetrating depth for which the magazine was formerly known. Other magazines, both print and online, also publish reviews from time to time, and you should seek them out, add them to your publisher's list, and hope for the best.

Why did we say that? Because these days, reviews—especially those in "out-of-the-mainstream" fly-fishing magazines—are sometimes the work of people who lack much knowledge of fly-fishing literature, history, or the English language itself. They also frequently misunderstand the purpose of book reviews and how to write them.

Just what is their purpose? A reviewer's primary job is to serve as a consumer advocate, trying to assess each new fly-fishing title in terms of its value or interest to potential readers so they can decide whether to invest in the new book. In other words, a reviewer's basic job is to tell readers whether a new book is any good.

But it goes far beyond that. Determining a new book's worth also means sizing up the quality and quantity of the information it contains; its newness, if any; the quality of its prose and illustrations, and its place, if it has one, in the great centuries-old panoply of fly-fishing literature. That's a tall order, and it's why reviewers of fly-fishing books should be well-versed in both the history *and* literature of angling, as well as the modern currents and ethics of the sport. They should also be familiar enough with English usage to be able to recognize both the high and low points of an author's prose. Apparently it's difficult to find people who meet those requirements these days, which might be another reason why we now see fewer reviews in print.

A reviewer's job also is often misunderstood. Just who in the hell appointed *that* guy to be the judge and jury of your new book? And what gives him or her the right to do so? Perhaps the best way to answer those questions is to compare a book reviewer's task with the process of scholarly peer review. Before scholars in many academic fields publish the results of research or announce a new discovery, they routinely send their manuscripts to a book publisher or professional journal, which in turn sends them to other experts in the field who evaluate the quality of the research or try to replicate its findings. Their collective verdict largely determines the value of the author's work, whether it even merits publication, or if it requires additional polishing by the author. The editor of the journal or book publisher then shares those comments with the author, who must respond to the editor's satisfaction before publication can proceed. (Paul has done a lot of scholarly writing in fields other than fly fishing, and was always grateful beyond words for these candid, competent reviews, although sometimes his skin wasn't quite thick enough.) Once the work is published, it is subject to further scrutiny in the form of reviews in professional journals. In this way, a scholar's work is reviewed by his or her peers twice, once privately and once publicly, which is one reason such publications have a higher general credibility than popular publications.

Music, literature, film and visual art adhere to a somewhat similar process, where each new work is judged by critics who apply the benefit of their experience and knowledge to reach an opinion, and it's the consensus of these opinions that largely determines the worth of the work. It's not a perfect system and usually there's no way an offended author can appeal the collective verdict, but it's the method society has evolved to establish a reasonably fair measure of the value of artistic things. And the critic's part is essential.

Some modern publishers of popular books, including fly fishing, similarly send a promising manuscript to someone whose authority they trust before deciding whether to publish it, but essentially all the reviewing a fly-fishing book gets will come after it's in print.

In some ways, a book reviewer's task also is similar to that of a baseball umpire: He or she must remain strictly nonpartisan and call 'em as he or she sees 'em. That means pointing out both the good and the bad—or, sometimes, the downright ridiculous. Steve, who wrote book reviews for 35 years and was a member of the National Book Critics Circle, remembers some of those ridiculous moments, like the writer who described a day when the fishing was slow and he resorted to trolling flies on two rods simultaneously. Curious about the regulations pertaining to the water the writer was fishing, Steve looked them up and discovered it was illegal to use two rods in that water, and said so in his review—much to the discomfiture of the author. Surely the writer of a fly-fishing book should be aware of the regulations where he fishes! Another writer, attempting to describe the behavior of birds during a hatch, got his species mixed up and said it was sparrows feeding on the hatch, not swallows. The same author also apparently had trouble with postal abbreviations; he listed several famous Alaskan rivers as being located in Arkansas!

Mistakes like this can be fairly humorous, unless you've just spent good money for the book (or unless you're the one who *wrote* the book). Sometimes, however, the errors can be more serious. One author boasted credit for making a new entomological discovery, but Steve had a nagging memory he'd read about the "discovery" somewhere else, and after a long search found the book where he'd read it—published more than a half-century earlier. The resulting review was unfavorable, to put it mildly.

Even if such errors are obvious, good reviewers will not let them get out of proportion but instead try to find a way to mention them in an otherwise laudatory review, or just treat them as minor matters that can be fixed with the book's next printing. There's an old saying among reviewers that if you find yourself having to criticize the book's index, maybe you just need to admit that it's a good book. Of course if there are so many errors that they compromise the book's usefulness, the reviewer must say so. Paul has reviewed books so rife with misconceptions and factual errors that he had no choice except to at least summarize them; potential readers deserve to know. He has found it especially difficult to handle such erroneous texts—and has said so in published reviews—with books whose general message and intentions he admired.

Authors should never count on publishers or copy editors to check this sort of thing; they might, but ultimately it's your responsibility to police your own stories.

It's never a reviewer's job just to blindly praise a new work; a good reviewer must also challenge the author's assertions to make sure they are well supported and accurate. If those assertions fall short, and the review says so, the author's reaction is likely to be angry, leaving both reviewer and author unhappy. Having occasionally been the target of negative reviews himself—sometimes he thought fairly and sometimes not—Steve knows how much it hurts. But that's good experience for a reviewer to have; they should all know what it feels like to be a target as well as an author of reviews. Unfortunately, too few of today's reviewers seem to have had such experience.

Some fly-fishing magazines follow the custom of distributing their reviews among a stable of different writers. That can provide a welcome diversity of opinions, but it also sometimes leads to a serious lack of consistency. Which method is better? Take your choice.

If your new book gets a favorable review in any media, you can hardly hope for any better publicity. But the opposite is true if it receives a bad review. If that happens, is there anything you can do about it? You can always write an angry letter to the reviewer or the magazine, blog, or podcast that published the negative review, and that might help lower your blood pressure, but it's generally considered bad form in the literary community to do that. Sure, it hurts to be criticized, especially if the criticism is unjustified, but the best thing to do is suffer in silence and hope the bad review will soon be forgotten by everyone except yourself. Viewing it from the most mercenary perspective, your published response will only increase the number of readers who are aware of the negative review, which won't help sales.

When Your Book Is in Print

Once your book is in print the next step is to work as hard as you can to sell it. Our first bit of advice on this subject is that you probably don't want to try selling your own book, even if your publishing contract allows you to do so (some contracts don't). Why? Just think about what's involved: You'll need to maintain your own inventory of books, which means buying them from the publisher at a discount, paying the shipping costs, lugging them to your home or office, processing orders, taking care of all the associated accounting, packaging the books you sell, then paying postage or shipping costs to send them to buyers. You really want to do all that? (See the next chapter for more about this.) It's a lot easier just to rely on the system that publishers, distributors, wholesalers, bookstores and internet booksellers have contrived to take care of the whole process. Anyway, you'll be busy enough just trying to continue promoting the book; it only makes good sense to leave the sales process to the professionals.

That process, however, remains a mystery to most writers, and sometimes it's difficult for new authors to leave sales of their book in the hands of a system that functions mostly behind the scenes and involves just about everybody in the publishing industry. That includes you, even if you didn't realize it.

At the top of the system are the mega-publishers and conglomerates that now dominate the book industry; in their shadow are smaller independent, regional and specialty publishers, and at the far end are the lonely self-publishers, though their numbers are increasing steadily.

The process begins well before your book is even off the press. It starts with the publisher, who is responsible for producing the book and, to a certain extent, promoting it. Most large publishers publish print catalogs of forthcoming titles and post them online to alert the public and potential customers that the book is forthcoming. Large publishers also employ or contract with salespeople whose job is to sell books to distributors, wholesalers and chain bookstores. Most large publishers hold two sales conferences annually, one shortly before the Christmas season to take advantage of holiday sales, the other in May or June, before onset of the summer "book season." Each conference gives the publisher an opportunity to introduce its new titles to the sales force, especially those the publisher plans to emphasize—potential bestsellers and the like.

It's not likely the latter will include many fly-fishing books.

The salespeople then begin meeting with representatives of book-distributing companies, which provide warehousing and shipping services and employ or contract with their own salespeople, who fan out, either in person or electronically, to meet with large potential buyers, such as libraries, book clubs and big independent or chain bookstores, and begin taking orders for the new titles. These negotiated orders specify the number of books each customer wishes to buy and the retail sales commission. Once orders are obtained, the salespeople pass them to the book distributor or the publisher for billing and shipping. This also gives the publisher an opportunity to assess the volume of orders and, if necessary, adjust the size of a book's press run in an effort to produce the correct number of copies for distribution.

Book distributors also contract with publishers to sell books on consignment, usually paying the publisher 40 percent of the book's list price. They also often work with smaller, independent publishers unable to employ their own sales forces, usually insisting on exclusive territorial distribution rights. If a bookstore customer fails to sell its full consignment of a particular title, the unsold books may be returned to the publisher for credit. Such books then are often recycled and sold in "remainder" sales at a steeply discounted price.

Book clubs sometimes buy a portion of a publisher's first printing of a new book, or print copies of their own and sell them at prices lower than the publisher's retail price for the book, paying a small royalty to the publisher for each book sold to a book-club member. The royalty is usually shared with the book's author, which leaves little money left for the publisher, but book-club sales are still considered a good deal because they involve little investment or risk by the publisher and sometimes generate significant publicity that adds to the publisher's sale of the book at full retail price. Alas, gone are the days of the great outdoor book clubs; selection by the long-popular Outdoor Life Book Club could give a book's sales and notoriety a huge boost.

So-called premium sales also can enter into the book-distribution picture. These happen when a company buys copies of a new book to distribute as promotional gifts to customers. The books are sold by the publisher at a significant discount and cannot be returned. This usually doesn't happen with fly-fishing books, but a chapter from one of Steve's books was selected for recording by a professional narrator on a compact disc for distribution to customers of a company that manufactures outdoor gear. The CD turned out beautifully, but in the end the company backed out of the deal.

Other book outlets include supermarkets, drugstores, airport newsstands and specialty stores such as fly shops. Most such outlets mainly stock paperbacks, but some, especially specialty stores, also keep hardbacks on their shelves. Publishers regard the latter as productive places to display and sell "niche" books, such as fly-fishing books, which are sold in many fly shops.

That's how the system usually works for the big publishing houses. Smaller or independent publishers often bypass distributors and work instead with wholesalers, who maintain large stocks of books in all categories and depend on catalogs and publicity to market them directly to bookstores, libraries or other outlets. The wholesalers in effect function as giant bookstores themselves. They ship orders and bill their customers directly. They do not employ their own sales force.

Sometimes the difference between distributors and wholesalers is a bit fuzzy; some distributors also function as wholesalers, and vice versa.

This is actually a very abbreviated description of the sales process. But before any book can be sold, people first have to know it exists, and that's where you, as the author, have an increasingly varied and significant role. Here's how:

Readings and Signings

Bookstore readings and signings are excellent ways to generate publicity and sales for a new book, especially in your home community. That remains true even in this age of digital communications and so-called social media. Most publishers used to arrange such readings for their authors, contacting bookstores to get the new author on the store's event calendar, coordinating the events with the author, even producing publicity materials such as photos or window posters, and encouraging the store to advertise the event in local news media. All the author had to do was show up for the reading.

Some publishers still do that, and we hope your publisher is among them. But in this age of huge publishing conglomerates, which deal simultaneously with dozens or even scores of new books and authors, some have gotten into the practice of leaving it up to authors to arrange their own bookstore appearances. This is especially likely with niche books, such as those pertaining to fly fishing. It also leaves authors in the very awkward position of having to phone or email bookstore event coordinators and try to convince them that their new book is worth a spot on the store's event calendar. Since some event coordinators, especially at large or chain bookstores, remain accustomed to hearing directly from publishers' publicists, their likely response to such calls is to conclude the author has been abandoned by his or her publisher, or the book is aimed at too narrow a market and is therefore unworthy of a place on the store's event schedule. You can argue, but there's no appeal, and you're probably going to be left out in the cold.

Not all bookstores will turn you down, however; some pride themselves on promoting local and regional books. Rare is the Barnes & Noble that doesn't have a dedicated set of shelves for local authors and local subjects, and Paul has noticed several of their stores advertising readings and signings by local authors whose books will almost certainly have only local appeal and may even have been self-published. These stores have recognized that behaving like a caring member of their local literary community can be good business.

Smaller bookstores also are more likely to welcome hearing from authors trying on their own to get on the store's events calendar. And if you're one of those authors, always remember that if you think you're trying to do a favor for the bookstore, the bookstore also is doing one for you—so don't even *think* about trying to charge a fee for your reading.

Fly shops also sometimes host readings and signings, and if you have to make your own appointments you're likely to receive a friendly reception from them, especially if yours is a "how-to" book. And you'll probably be welcomed with open arms at local fly-fishing clubs. More about that a little later.

No matter where readings take place, they can be enjoyable social occasions because you never know who might show up.

Among Steve's readings was one attended by a high-school classmate he hadn't seen in decades, another by a fraternity brother not seen since college days, still another by an old favorite college professor. Even if you don't know anyone in attendance, you'll likely meet a friendly, inquisitive audience that will ask some good, thoughtful questions. So enjoy yourself.

On the other hand, readings and signings can sometimes turn out to be painful occasions. You might sit there for an hour waiting for people to show up, and nobody does, which probably means the bookstore and/or your publisher did nothing to promote your appearance. So you sign a few books for the store proprietor, just to cover your embarrassment and his or hers (remembering that signed books can't be returned to the publisher), and go out for a couple of beers.

Also keep in mind that lack of attendance doesn't always necessarily mean bad news for the bookstore owner. The proprietor of Paul's favorite bookstore in his former home in Montana, where he did a number of signings and readings, told him he should never worry about the size of the crowd because the bookstore's advertising in advance of the event always paid off; many people noticed the ad and over the course of a few weeks following the event they would come in and buy the book. For that reason, the store manager always had Paul sign all the copies ordered for the reading, because she knew they would sell eventually.

Here's how to prepare for bookstore signings/readings: Let the publisher make the arrangements; don't try to do it yourself unless you have to. The publisher should encourage the store to promote the reading aggressively in local media, by printing and distributing posters (if you can get hold of one of the posters, it will look great on your office wall), planting items in local newspapers, etc. If the publisher isn't involved, then the bookstore should advertise.

Choose what you plan to read, preferably no more than a half hour in length, and make sure it's something that doesn't leave the audience hanging in the absence of some sort of conclusion. If you can, it's a good idea to choose something familiar to the audience—a local water, local angler, local history, etc. Start by explaining a little about how you came to write the book and what you hope it will accomplish. Rehearse your reading out loud until you're satisfied the nuances and voice inflections are what you want. It's not unusual to hear people attending bookstore readings say your prose "sounds" different when they hear you read it. Rehearse several times if necessary.

Get to the bookstore early to check out the venue and make sure everything is satisfactory, especially the sound system. This can avert possible disasters. At one bookstore where Steve was reading, the microphone stand had a loose fixture and the microphone kept slipping lower and lower until he finally had to detach it from the stand and hold it in one hand while trying to hold the book in the other. Then he lost his grip on the book and it fell to the floor with a loud slap. Things went downhill rapidly after that.

For signings, you'll want to have a couple of pens you know will work and you should have in mind some previously made-up phrases or greetings you can use at the spur of the moment so you don't have to try to think of something appropriate to write in a book. Try to avoid using the same ones too often, however; sometimes people will compare books to see what you wrote in each of them. Most important, if someone wants you to autograph a book for a specific person, make sure you ask how the person's name is spelled. Nothing is more embarrassing than autographing a book to someone whose name is misspelled (it's also an expensive waste of a book).

Have fun with this; don't get into a rut of "with best wishes" and "Tight lines!" and other worn-out phrases. If you know the person, consider going for a laugh: "To my pal Ferd, who taught me everything I know," or "Bob: Don't bother checking the index—you're not there." Even with strangers you can often add a bit of a laugh: "Nigel: Thanks for showing up for my reading; I was afraid nobody would!" We realize this approach is a bit risky, but usually people will walk away happier if you've given your inscription a personal touch.

Some venues will be much better than others. Steve discovered the Seattle suburb of Lake Forest Park offered one pretty close to ideal: One of the city's best bookstores, Third Place Books, is on the top floor of a mall, and the Avid Angler fly shop is on the ground floor—a perfect opportunity for joint promotions involving both stores.

Another store where Steve read offered free glasses of wine for all in attendance. The audience was the largest he ever had for a reading, and he suspects it was the wine, not his presence, which drew the crowd, but who was complaining? And even though the wine was free, he didn't drink any; it's always best to abstain before a reading so you don't start slurring words; afterward, however, is OK—and often warranted.

As mentioned previously, local fly-fishing clubs are always on the lookout for monthly programs, and if you belong to one all you need do is give the program chairperson a call. You can find other club contacts on the internet and call them, too, although the best circumstance is if they come to you first—and they will, once you establish a reputation as a fly-fishing author. You can prepare for each appearance as you would for a bookstore reading, although since you're dealing with an audience of confirmed fly fishers you may want to expand your program, perhaps even talk about something besides your new book. In any case, you should prepare carefully for each appearance, tailoring your presentation to the local audience as much as possible. Rehearse out loud just as you would for a bookstore reading until you're certain you've got your spiel down pat. This always works better, even if you're a competent extemporaneous speaker. Having a video, slideshow, PowerPoint presentation, etc., can make it easier, but remember that using such visual technologies poses its own risks and don't jump into it unless you're prepared to get good at it before taking it on the road.

Find out beforehand if your presentation will be at a dinner meeting; if not, better grab a burger ahead of time. Also ask what to wear; some clubs are more formal than others. As a rule of thumb, men should wear a jacket and tie; those can always be removed if nobody else is dressed similarly (as men, we hesitate to offer any advice on what women should wear to a fly-fishing club meeting; your judgment is bound to be better than ours).

When you put on a program for a fly-fishing club there's a very good chance you'll be at the end of the evening's agenda, so be prepared to sit through fishing reports, club business, and other trivia before they finally get around to you. But such meetings also can be a lot of fun; you'll probably hear a few good stories, maybe some of them even true, and if you buy a few tickets for the evening raffle you might win something worthwhile. One meeting Steve attended also featured an auction where he was able to purchase some excellent fly-tying hooks at a real bargain price.

Should you charge a fee for a fly-club presentation? That's up to you; most clubs will offer one. But remember, fly clubs are non-profit organizations, usually with limited budgets, and they don't have deep pockets to pay for programs. Steve's policy was to ask only for reimbursement of expenses and waive any speaking fee. Why? He didn't need the money, and anyway he thought that if he had a bad night and left a disappointed audience, he wouldn't want to leave with any of their money in his pocket under those circumstances, especially if it meant the money could have been used by the club for other, more meaningful purposes. Besides, in a way the club was doing him a favor by letting him publicize one of his books. That policy obviously wouldn't work for everyone, but it worked very well for Steve.

Here's another tip, from an old folk-singing friend: If you develop hoarseness from reading or speaking (as the folk singer did occasionally while singing), brew some Throat Coat tea in advance, refrigerate it, then take it with you in a thermos to the reading and have a few swallows if you start getting hoarse. It'll help immediately. Be careful, though; if you consume too much, it could loosen up other things.

Podcasts, Blogs, and Fly-fishing Shows

Somehow, in the pre-internet past, fly fishers got along without podcasts and blogs. Some still do. But whether you like them or not, these digital delights are prime targets for promoting sales of your new book. That's because at least some podcasts are always anxious to interview authors of new fly-fishing titles, and blogs also are useful for keeping your book in the public eye.

That's not to say that either podcasts or blogs are noted for professionalism or accuracy; often they are the work of individuals lacking experience or education in either audio or visual media, and some also seem without much experience in use of the English language. So if you choose to publicize your book in either of these ways, be prepared to look the other way when the embarrassing grammatical errors start to flow. But a little embarrassment may be worth it; your publisher thinks podcasts and blogs are good ways to publicize your book.

On the other hand, if it comes down to a choice between listening to a podcast or going fishing, the latter might be your best choice.

If you search the internet for fly-fishing podcasts, you might be surprised at how many you find, and most are easily accessible through a variety of common digital devices—assuming the devices have earphones or speakers. Some podcasts are sponsored by companies whose names you will probably recognize, others by individuals whose names you will likely forget after listening to them once. A couple of podcasts reminded Steve of the high-school class where the teacher said "uh" so often between words that the students organized a daily pool, trying to guess the number of "uhs" she would utter each day. Steve kept score until the bell rang and the winning total could be announced.

Some podcasts are like that. But if a podcast host or sponsor calls or emails and asks you to sit for an interview about your new book, by all means go for it. It's likely you won't even have to go anywhere; just sit in front of your computer with its camera and microphone turned on and answer the host's questions. They will probably include such queries as why you wrote the book, how you wrote it, how long it took you to write it, and a few words about what's in it. Most podcasts are still audio-only, but more and more seem to have video too. In that case, it will help visually if your computer's camera is trained on the wall of a book-lined study behind you, or a wall covered with framed fly plates or fishing photos. You also might want to dress as if you had just left the stream, although waders probably aren't necessary or advisable.

Podcasts, it appears, are here to stay. And as the old saying goes, there's no such thing as bad publicity.

Blogs are a different matter. They rely mostly on email "posts" from viewers responding to a question or offering viewpoints on a common theme. These are usually displayed in reverse order, so you have to go back to the beginning to learn the original question or theme. This can be pretty frustrating if lots of people are signing in and offering answers or opinions, and sometimes the discussions continue for a long time—even years. Blogs also are places where you can often find some of the most outrageous transgressions of good grammar and spelling you ever imagined. However, they are also places where you can get some publicity for your book, perhaps just by mentioning it subtly in answer to a posted question.

As with podcasts, if you search the internet for fly-fishing blogs you'll probably discover more than you ever knew existed. Again you'll find some sponsored by companies with familiar names, others hosted by fly-fishing magazines, fly shops, guide services, travel services and more, from all over the world. Blogging is definitely an international sport.

Some blogs appear to have morphed into online magazines (and some continue in both realms). Some are also purely advertising sites. In fact, there really is no precise definition of the term; it seems a blog can be whatever its masters want it to be. If it advertises itself as a "fly-fishing blog," that narrows it down a bit, but as we all know, fly fishing is a topic with an almost infinite number of threads.

Which also makes blogs good places to publicize your book. However, it's advisable to explore at least a few blogs before you decide to post anything on one or more of them; remember, you may be judged by the company you keep.

If there is an annual commercial fly-fishing show in your vicinity, that also can offer a good opportunity to publicize your book. Find out if the show has an author's booth; if so, sign up for a time slot. This can be just for signing books, or, if the show sponsor (and your publisher) will allow it, it also can be a chance to sell books, if you want to try doing that yourself.

Book Tours, Internet Reviews, and Websites

We've all heard about book tours where celebrity authors are squired around on private jets and limousines, stay in the finest hotels, and get lots of face time with television and print media. Don't expect that to happen to you; even publishers with the deepest pockets aren't going to spend that amount of money sponsoring festive tours for their fly-fishing authors. If it *does* happen to you—well, consider yourself lucky.

Steve was the fortunate recipient of one "tour," though it was confined to a single city, took only two days, and he had to drive himself to get there. He was quartered overnight in a hotel that might once have been the city's finest but had since become probably its oldest, and the surrounding neighborhood had declined along with it to the point where one's personal safety was at risk. When the publisher's assistant arrived at the hotel to pick up Steve, she wasn't a glamorous creature riding in a chauffeured

Rolls-Royce; she wore jeans and was driving a Jeep Wrangler. But she was good company and got Steve to his appointments with television, radio and newspaper interviews, and overall it was an enjoyable experience.

Another quick and easy way to promote sales of your book is to encourage people to post quicky reviews on websites such as Amazon or GoodReads. If you have ready access to the internet, you've certainly seen such "reviews," rating a book, product or service from poor to good on a scale of 1 to 5. Whenever you have an opportunity, even if it's with only one person or a large audience, tell them that if they like your book you would really appreciate it if they would post a favorable "review." Just a single line—something like "Joe's book is the best one yet on how to tie furled leaders"—along with a "5" rating will do the job. Believe it or not, publishers love to see these and they do go a long way toward promoting book sales.

Some publishers think personal websites are another good way to publicize books and promote their sales, apparently on the theory that if people search for you on the internet they'll find your website, read what it says about your new book, then maybe decide to buy a copy. That might be true, but there are two sides to the proposition: It takes a little time and money to establish and maintain your own website, and if you plan to use it to sell your book, it might be worthwhile, at least for a while. After weighing the pluses and minuses, Steve decided against it but Paul opted to establish one. As a member of the Authors Guild, he took advantage of its inexpensive service to set up and host a website for members. His goal—besides pacifying friends who kept bugging him that he should have a website—was to present information about his books in a way that would encourage people to check them out. So now the site lists his books, provides review quotes and extended excerpts, announces new books and awards and offers other notable information. If he ever decides he wants to have a blog (this is wildly unlikely), the website provides a ready format for that, too.

But it can take lots of time, effort and energy to maintain a website and keep it up to date, and long ago Paul made the fundamental decision that if it came down to a choice of either cultivating an audience for the website or writing his next book, the book would always win. Meanwhile, he can tell from the number of people who contact him via the website that at least some are finding their way to the website to do it, and to that extent it's serving its purpose.

While we're at it, the Authors Guild deserves further mention. It's a large, venerable professional society of writers, including not only obscure specialists like Paul but many of today's best-selling novelists and nonfiction writers. Because of the size and significance of its membership, the Guild has considerable political clout in lobbying for writers' rights, providing legal expertise to members, and otherwise being supportive of the writer's trade—including providing the service of setting up and maintaining websites. Now that you're an author, it's worth your time to check it out at https://go.authorsguild.org.

It is obvious that reader and author share in a book's success and that the character of that success depends on both of them.
—ARTHUR RANSOME, *MAINLY ABOUT FISHING*

CHAPTER TWELVE

Should I Self-Publish My Book?

THERE WAS A TIME IN THE PUBLISHING BUSINESS when most informed people perceived an author's decision to self-publish a book as nothing less than admitting defeat. Self-publishing meant that you had utterly failed to find a publisher foolish enough to commit the necessary investment in review, editing, design, printing and marketing that went into a "real" book. At that point, so this reasoning went, your only remaining choice was either to abandon the manuscript or pay for its publication yourself.

That's a personal choice, and it's not for everyone. Some authors might feel it's a form of surrender. Late in his career, Paul's friend Ken Cameron, gifted and versatile author of dozens of historical novels, detective novels, espionage thrillers, college textbooks, and many shorter works on the culture and history of fly fishing, took a hard look at the possibilities for self-publishing some of his new novels, as well as self-publishing new editions of his earlier works, but decided he just couldn't do it. Ken, who was published by many large presses in the United States and the UK, probably spoke for a great many authors when he decided that even as a well-respected and profusely published author in several genres, he still needed the "validation" of having his books accepted by a legitimate press that would bear responsibility for publishing and selling them. It's true that validation is central to the author's art, and if you don't get it you may have lost much of the reason for writing anyway.

But self-publishing has changed hugely in the past 20 years. It is often very attractive, and can be the right thing to do in many circumstances.

How It Was and How It Is

Until the digital age, the simplest way to self-publish was to hire a local printer, give them your manuscript, and trust them to put it out in good time and in an appealing form. That approach worked well in countless cases, say for example someone whose goal was to write a little history of their family, their town, or their church. One reason it worked so well for these projects was that the author wasn't especially concerned with making money so much as preserving and presenting a story or a lot of information for a clearly defined and usually quite small audience. The author typically knew that the print run would be quite modest anyway, no larger than the number of family members, church members, or town citizens who might want a copy.

Obviously, the more money you could invest in this production, the more professional would be the result. Given an author with deep enough pockets, it has never been difficult to hire a classy printer—or even to contract with a professional publisher—who would spend your money generously on a competent editor and proofreader, a designer with good taste, and the various suppliers who provide high-quality paper, binding, and other services necessary to make a book at least as handsome and appealing as any book produced commercially by the big industrial publishers. Historically, some of fly fishing's most beautiful and even distinguished books came into being that way, usually in fairly limited editions. Some of those limited editions, the ones for which there was a larger audience out there, even had second lives as cheaper "trade editions" picked up by commercial publishers after the original book had been out for some time.

There also were so-called vanity publishers, whose books had all the outward appearance of books produced by traditional publishers, but who in fact charged substantial fees to produce and promote the author's book. Here's a single sad example. For a few decades, the largest vanity press in the United States was probably Vantage Press, in existence from 1949 to 2002, when, following a Federal Trade Commission investigation, the New York State Supreme Court ruled that it had defrauded more than 2,000 authors who did not receive all the professional services the press promised them. But in the meantime, Vantage had at least provided those authors with books that looked as "real" as any book published by traditional presses. No doubt a great many authors were reasonably content with their books. As well, serious book shoppers, seeing the "Vantage" imprint right there on the book's spine, could hardly be blamed for not knowing that the book might not have been quite real in the traditional sense. It couldn't have hurt the seeming credibility of the book that there was, confusingly, also a prominent and respected large publisher's imprint named "Vintage."

The controversy and publicity surrounding the exposure of Vantage's unsavory doings can only have reinforced the image of self-publishing as something beneath the dignity of legitimate writers. This is especially sad in the case of the Vantage books that might well have been quite good but could never escape the taint of vanity publishing. The validation, at least for potential readers, just wasn't there.

It is revealing of the good-versus-bad aspects of this sort of publishing, however, to repeat that Vantage did a tolerable job for a lot of authors' books, including at least one about fly fishing. In 1972 it was Vantage that published Charles Kroll's excellent *Squaretail*, which was for some years the essential book about brook trout. We know nothing of how well *Squaretail* sold, but we both remember that at the time it was well regarded. By whatever marketing or simple good luck it enjoyed, it received substantial attention among anglers (as far as we know, it had no subsequent editions, either from Vantage or from being picked up by another press).

The Vantage saga is hardly the whole story. In the long, convoluted history of publishing, there have been many experiments with the idea of how to make a book. If you look hard enough at those different approaches, you can find "proof" of any approach being the best, and equal proof of that same approach being a dismal failure. It depends upon many things, some of which are the inherent intangibles that every publisher faces with almost every book it decides to produce.

However, there have been plenty of circumstances in which self-publishing worked so well that there was really no point in even considering a conventional publisher. A few examples from the world of fly fishing explain why.

First, Steve, a longtime member of the Washington Fly Fishing Club, one of the most important early West Coast fishing organizations, volunteered to write that club's history to help celebrate its 50th anniversary. It appeared as *Backcasts: A History of the Washington Fly Fishing Club, 1939-1989*, published in the latter year by the club. Like hundreds of other club histories published during the past couple of centuries in the UK and North America, it was produced in a style, format and number sufficient for the needs of the membership, with no intention of getting into the larger, general-audience market of fly fishers. It served its purpose, and was thus a perfect candidate for self-publishing, which at the time might have been the only practicable route for getting a book with such a limited audience into print.

It's worth knowing, however, that today, as the history, culture and business of fly fishing have grown in credibility as significant research topics even among some academics, it might be possible to interest either a regional commercial press or even a university press in a title like this. As we saw in chapter 10, since 1976, when the globally prestigious University of Chicago Press published *A River Runs Through It*, a number of other university presses have published fly-fishing titles.

Paul's first experience of this sort was in the cowriting of a corporate history, yet another type of publishing that exists in its own right, often far separated from the mainstream of commercial publishing. In Paul's case, the book was *The Orvis Story*, which he cowrote with fishing historian Austin Hogan in 1981 to celebrate the 125th anniversary of that grand old tackle company. This first edition was small enough and, frankly, enough fun, that Paul wasn't concerned about getting paid much and happily accepted most of his payment in some great Orvis gear that he has used ever since. Orvis hired a talented local designer there in Vermont to work with Paul on selecting the illustrations and otherwise seeing the book through the design and printing process. Once printed, it was offered for sale in the Orvis store and catalog, which put it in the sights of many thousands of sportsmen, who bought out the whole press run. It was, in effect, a promotional tool, but one with considerable accurate historical reference value as well.

Paul's good experience was repeated 25 years later (as the song says, who knows where the time goes?), when Orvis invited him to completely revise and greatly enlarge that book as sole author, to celebrate its *150th* anniversary. By this time, much more research had been done into the company's history, so it was exciting to be able to tell the story much more fully. As important, by 2006 Orvis was a much larger company, so the book came out proportionately dressier and elaborate, lavishly full-color throughout, with a beautiful slipcased limited edition in addition to the regular trade edition. Orvis distributed complimentary copies of the books to appropriate people, and again offered it in its catalog and its by-then sizeable number of large stores and shops around the country. It sold well and served its purpose. (Paul got paid a lot more this time, too.)

A telling aspect of this second edition was that by 2006, when it was published and the company was so large and such a venerable and influential part not only of fly fishing's history but of American business history, the book could have been of interest to one of the outdoor-book publishers or even a general trade publisher. But Orvis chose—and Paul thinks wisely—to go it alone and maintain complete control over all aspects of the book's production, not the least being free to get it out on its own schedule. As we've mentioned before, there can be a very long lag, as much as two, three or even four years, between the time a book is accepted by a publisher and its final appearance on bookstore shelves. Using its own considerable in-house editorial skills, and bringing back the same talented designer/production manager who did the original edition, Orvis and Paul got exactly the book they wanted when they wanted it, and without having to share any editorial or production decisions with the corresponding editorial, design and marketing departments of a publisher. And, just in case you're wondering, no one at Orvis ever tried to influence Paul to "rewrite history" or put Orvis in any more favorable light than it already deserved in the history of the sport. That's corporate confidence for you.

A third helpful example of how well self-publishing a fly-fishing book can work comes from the unlikely little town of

Waseca, Minnesota. There, in 1941, tackle manufacturer and master mail-order salesman George Leonard Herter self-published the first edition of his *Professional Fly Tying, Spinning and Tackle Making Manual and Manufacturer's Guide*. Older sportsmen will remember Herter's company as the proto-Cabela's that generations of us found so important for providing us with outdoor sport's necessities from the 1930s to the 1970s. Through those many years, the ever-larger Herter's catalog arrived with a ponderous thump in countless mailboxes, generating a flood of orders from all corners of the sportsman's world.

It says much about Herter's commercial reach that except for his small-town store, his very popular catalog was in fact the only significant sales outlet that he had for his book. But it was enough. Bolstered by testimonials from famous people of the day (including Ted Trueblood and J. Edgar Hoover!), and prominently marketed in his phone-book-size catalog, the *Guide* sold so well that Herter could revise, enlarge, and reprint it every couple years. In 1953, only twelve years after its original appearance, Herter claimed a total sales of 400,000 copies—a terrific number even for a best-selling novel, much less for a book on a subject as specialized as fly tying.

Four hundred thousand? Truly? Who knows? Herter was such a notorious self-promoter and creative BS artist that few careful observers would trust that number. But the mere fact of so many reprintings, even if each had a fairly modest press run, is proof of an enviably successful venture in self-publishing. When Paul bought his now-ragged copy of the 1971 edition, it was the "revised 19th edition," an exhaustive, thoroughly illustrated 584 pages of fine print, black-and-white and color illustrations, startling opinions almost beyond counting, and vast amounts of undeniably valuable information. There could ever have been only one George Leonard Herter. We should be glad he came along, and maybe one was enough but, say what we will about him, he sure knew how to sell his books.

Beyond the entertaining and sometimes instructive details of the stories of these three self-published books—a club history, a corporate history, and an exhaustive fly-tying manual—there is one common and overwhelming theme here that is the same one we saw in the earlier, smaller-scale examples: their readerships were sure things.

And that's the rub of self-publishing; for every success story like these, there are many tales of authors winding up helpless victims of the great self-publishing downer, the garage full of unsold and unsellable books. Don't ever forget this. Innumerable writers who were just sure their books would sell forever ended up with stacks of dusty boxes of unsold books crowding the lawn mower or even the car out of the garage. Their books became literary mulch.

Staying out of the Garage

Modern self-publishing is a product of the same technological and digital revolution that enables us to get and read copies of countless old books, as described back in chapter 2. At the heart of this side of the revolution is print-on-demand (POD) technology, itself a subject worthy of your reading time. For our purposes here, it can be summarized much like we summarized the OCR scanning of books that makes them digitally available. There are now some amazing contraptions—the most famous are known as Espresso Book Machines (EBM), and you should Google that name for videos of them at work—that combine every stage of bookmaking into one VW Beetle–size device that includes computer terminal, laser jet printer, and bookbinder all in one. In precise effect, you upload a PDF of your book into one end and within minutes the thing spits a finished paperback copy of your book out the other end. Now that the technology has been created and refined for a while, the resulting book is no less handsome, durable, and professional-looking than any you might buy in the snootiest bookstore.

The advantages of this technology are several.

For one thing, it's more or less instant. Once you've prepared the PDF and it has been uploaded into the EBM, you are only minutes from holding the finished book in your hands (alert: the previous sentence is true, but it does oversimplify the "instantness" of the preparation, which we will discuss more in a minute). Take it from authors who have waited an inordinate amount of time for a publisher to get around to bringing out a book, this is priceless.

Paul remembers a book—a collection of Theodore Roosevelt's wilderness writings—that he sent off to a publisher who happily accepted it but then took nearly four years to bring it out. By then Paul had an agent, who saw to it that such a thing never happened to him again, but it is not uncommon for a publisher, for its own reasons having to do with the timing of its catalogs and other Ouija board–grade mysteries, to set a book's release date further and further (and even further) into the future or, much less often if it ever happens at all, to bring it out earlier than originally planned.

It's hard to overemphasize that traditional publishing has what many of us find to be a glacial pace. You send off the final version of the manuscript, and once accepted it goes through stages of editing, proofreading, illustration review and approval, design, and the coin-flipping process of deciding when it will be released—all of these chores being completed under the watchful eye of the people who put that publisher's printed and online catalog together and who typically work nearly a year ahead in scheduling books for appearance in those catalogs. Take it from us that after a few decades of waiting anywhere from nine months to three years to see an approved manuscript turn into an actual book, it can be no less than a giddy experience to upload your manuscript to one of the online platforms, order a copy to check over one more time, and within two or three days of uploading it, hold an actual FedEx'd copy of the book in your hands. As we'll explain in a moment, part of that wonderful immediacy of what seems like an instant book is illusory, but it must be said that uploading your manuscript on a Friday and then being express-shipped the finished book, *a real book,* the following Tuesday is a wonderful thing—at least until you start noticing the typos that escaped uncorrected, when you will go back into your computer file and correct them and then re-upload the whole thing. The POD publisher's computer receiving them seems not to care how many times you need to do that; after all, each time you're paying for the sample copy to look over, and for shipping them to you (Paul, not the sharpest of proofreaders, had to do this three times with one of his books before getting it cleaned up, but it still took only a couple weeks from the time of initial upload to the time of having the book right).

Another advantage of POD self-publishing is that for those writers who want absolute control over their book, this is the way to get it. Of course with that control comes responsibility for all the things that can get screwed up in the publishing process, but more on that momentarily too.

For another, and this is probably the most important, the lawn mower can stay. The garage remains book-free, because you have ever only as many books printed as you need at any given time. And if, say, you're selling them to local stores (at whatever "wholesale" price you choose), when the store places an order you can just get online with your POD publisher and have the requisite number of copies shipped directly to the store.

Getting the Book Ready to Print

As we've hinted, that summary oversimplifies both ends of the Espresso Book Machine's moment of glory when your book is made real and you stand there gaping at it in wonder. Before letting the EBM have its swift way with your manuscript you have to do a lot of work; in fact, you have to do all the work that a traditional publisher would have done prior to handing your book to a printer.

By bypassing a publisher, you bypass all the important things they do to get a book ready for printing. The first and maybe most important of those things is good editing. We sadly recognize that with the coming of the web has also come a glaring decline in the average quality of the "published" word; all too many websites are built and run by people with remarkable gifts for software management and design but no noticeable skills at clean, clear writing. As newspapers have found their circulations undercut by web alternatives, they've been forced to reduce staff, with the inadvertent effect of lowering their editorial standards as well.

Don't settle for those lowered standards. For really important reasons you want your book to read as clearly and cleanly as possible, and the only way to do that is to get professional help. We've praised professional editors and proofreaders before so we won't rerun that exercise here. An ironic good thing about the collapse of some parts of traditional publishing has been that many of those editors who were cut loose are still out there and still doing good work; they've just retooled and set up web-based editorial services. There are many of these; check around on the many writer's-help sites to learn more about any of them. Once you find a likely one, make sure to talk with the person who will be working on your manuscript, and if at all possible check his or her references. They don't necessarily have to know anything about fly fishing to make a significant difference in your prose, but they might need some tips on the jargon; putting them onto one of the fly-fishing magazines' style sheets could make a big difference. Last, right off, get used to paying them fairly; their rates are worth it.

Once your manuscript is cleaned up, if you're going to go the typical route of uploading it to an online print-on-demand company—Kindle Direct Publishing (Amazon.com), Barnes & Noble Press, Kobo, Lulu, Apple Books, and others are awaiting your business—you'll need to get it into final form as a PDF. This means you need to have the book's final appearance created using whatever book-design program is required by the POD's computer, which means that you or someone you persuade or hire knows how to do that. Since such design work is as sophisticated as any other profession, if you want your book to look professional you need to have it designed by a pro. If you're lucky, you know someone who does that sort of thing, but online design companies have proliferated along with all the other web services, and the same rules apply: look around, check references, talk it over with likely prospects.

With these editing and design stages, you often also have the option of just paying the POD company to do it all. The attraction of this is that the whole process of making the book, from manuscript to printed copies, happens in one electronic place, and there is probably less likelihood of glitches popping up as the book moves from one stage to the next. Many of them offer all sorts of package deals for editing, design, and marketing. We wouldn't presume to generalize about these offerings beyond the same advice of checking out the prospects, asking around and talking to the people you'll be working with. If possible buy two or three of the books they've done in the past, just to see how they came out. The big companies are undeniably very good at producing slick designs with best-seller-ish covers.

Selling It

If sales and readership matter to you, this is where you'll have a lot of very hard and not necessarily fun work facing you. For example, say you have Kindle Direct Publishing make your book. A happy part of the deal is that they will automatically upload it onto Amazon.com (for a modest additional fee they will also convert it into a Kindle book, which you'll probably want to do because e-books are such a significant part of books sales these days). But it is naive in the extreme to imagine that just because your new Kindle book is now handsomely available as a paperback and e-book on Amazon.com, anyone is going to buy it or even notice it among the 30 million other titles they offer. Unlike a book produced by a reputable commercial publisher, sales and promotion of your book are entirely up to you.

As with apparently everything to do with the modern digital/electronic world, there are countless websites, books, apps and other sources of information on how to do things related to book marketing and promotion. If you're serious about this, spend time tracking down the information that seems most to apply to your book and your hopes for it. All we want to do here is prepare you a bit for what you're in for. We don't mean to be dismissive or flippant about this, but the simplest way to explain what's involved in promoting and selling your book is for you to read chapter 11, then imagine yourself doing everything in it absolutely on your own, and then figure out which of those chores you're able and willing to hire out. If, as in some of the examples given above, you don't really need marketing help, or just don't care whether or not your book sells, the intense efforts described in chapter 11 probably don't even apply to you. But if you do care, and you urgently wish for as many readers as possible, then you're in a whole different realm of the publishing business. Ever write a press release, much less figure out where to send it? Do you enjoy building and regularly, if not daily, updating and feeding a website? How about developing a very large or at least very well-targeted email list of potential buyers, another list of where to send review copies and another list of local stores and fly shops that you'd like to convince to carry your book? By considering these and a number of other questions, we don't mean to discourage you, but your answers will probably tell you a lot about just how sure you are that you want to self-publish in the first place.

How It Finally Feels

In the past few years, after Paul had published 40-some books with small, medium, large, and giant publishers as well as half a dozen university presses, he realized that he was getting too old to worry much about needing more validation with every new book he wrote. He had no intention of abandoning the traditional publishers, who were still interested in at least some of his books, but perhaps because he did like the apparent independence POD publishing brought to getting a book done, he was curious enough about this new alternative to dabble in it. It has been a revelation. He hasn't yet published a fly-fishing book that way, but he does have a fly-fishing project or two in mind for which self-publishing may be the best way to go.

Besides all the good and not so good practical aspects of self-publishing, Paul is convinced there are intangibles to be considered. The first time he went through the process he wondered how a book made like this—produced without the technical support, professional validation, and the possible advance money and eventual royalties that go with working with a traditional publisher—would feel when he got that first copy in his hand. Authorship is full of personal, intangible moments like this, where some of the trade's richest rewards may be found. What he discovered that first time he received an actual copy of a self-published book from one of the online platforms, was that it felt pretty good. Really good, in fact. The book couldn't have looked more real, with all the heft and other simple physical attributes, as well as all the less definable promise of any book that cordially invites you into its pages to see what it might have to say.

It helped that the books he's self-published were done that way mostly because he wanted them to exist; he went into it fully aware sales would be tiny, especially considering the certainty that he wasn't going to bother with marketing them (for the simple reason that he'd rather be writing another book). It was very much like vanity publishing, really. His primary goal was simply to make the books real and available to the few people who might search the bookseller websites for books about his topics.

He has no regrets about this. He's now done a few books that way, some wholly new and some just newly revised editions of books that were originally published by a traditional publisher. The whole thing has been a hoot, in good part because he was lucky to have friends he could enlist to help him with design, typography or whatever else he was incompetent to do. His wife, the artist Marsha Karle, provided cover art and sometimes even illustrated the books.

But he can also tell you that no matter how sure you are that you don't care about sales, and only want the book to be out there in case someone wants it, there is going to come a day, or more likely a late evening, when you'll notice your proudly self-published book sitting there on the shelf and will find yourself thinking how nice it would be if it just had a few more readers.

When writers die, they become books, which is, after all, not too bad an incarnation.

—Jorge Luis Borges

Last Words

That about covers it, except for a few thoughts that we put last just to make extra sure you remember them. Writing a book, for all its joys and challenges, finally boils down to working hard at your own professionalism, and here are a few words about how to do that.

Throughout this book we've told quite a few cautionary tales about how things can go wrong in your dealings with all the other people involved in writing and publishing a book—editors, proofreaders, manuscript reviewers, designers, marketers, bookstore owners, and publishing houses. But we aren't trying to convince you all those people are enemies, and don't for a second imagine that you are all alone on some idealized literary high road. It could well be true that you care more than anyone else about your particular article or book, but never forget that everybody else involved in publishing your work has the same goal: to publish the best articles and books they can. There may be days when it feels like your publisher is just an obstacle between you and your readers, but such adversarial thinking will lead you nowhere. Do your best to see things from the point of view of all your partners in this great venture, and to keep on positive, constructive terms with them. It might help if rather than thinking of your potential publishers narrowly, as just another business, you see them as part of a good cause, that cause being the celebrating and advancing of fly fishing as a sport, a social enterprise, and a way of life.

The history of literature is full of stories of prima donnas—writers who somehow got the notion they were God's gift to the English language, and everyone else existed only to praise their masterpieces to the world. It will do your prospects for publishing no good if you get a reputation for whininess, tardiness, intractability, or treating others as lesser beings (you'd be amazed at how often this does happen). Once you've racked up a couple of Pulitzers and at least one Nobel Prize, then sure, feel free to be as snotty and pretentiously self-important as you'd like. But until then, be considerate, helpful, and positively enthusiastic about the great fly-fishing enterprise you've become a part of. Play nice.

Professionalism means many other things, some of which we've already touched on but which bear repeating: Honor deadlines and don't make your editor have to pester you about getting the work done on time. Pay close attention to whatever style sheet your editor gives you, and get all those grammatical, spelling, and other specifics right *before* you send off the manuscript. Not only is that simply the right thing to do, but also the less time you make the editor waste cleaning up your manuscript, the more time he or she can spend on the things that matter, and the more kindly he or she will think of you next time.

When your manuscript is undergoing line-by-line copy editing, and it's returned to you with a fair number of minor editorial suggestions—deleted commas, minor wording changes, that sort of thing—avoid defensiveness at all costs. Remember all the good things we've already said about professional editors as a higher form of literary being, and look hard to see what you might learn. In more Machiavellian tactical terms, try to stand back and read each revision from an objective distance. There are lots of ways to say the same thing well. If a revision hasn't changed your meaning and the revision is a reasonable alternative to what you wrote, let it go. As we said earlier in this book, the more of those minor points you concede the better your position will be in getting your way with the few potentially larger changes you feel bound to resist. Choosing your battles carefully at this stage will save both you and the editor a lot of time and tension, and can generate some goodwill. ("He's easy to work with," rather than, "Jeez, I dread editing that jerk!")

What we're talking about here is as much about good manners as about the technicalities of who is right, or at least righter. Get in the habit of thanking people. For example, make sure your editor conveys your thanks to the copy editor or proofreader whose work has made a good difference.

Your relationship with your publisher and all the other people involved in making your dream into a book doesn't end when the books arrive. With any luck at all there will be work to do promoting the book and otherwise keeping it in the public eye. But before that next stage of being an author gets underway, take the time to bring the acknowledgments in your book to life. Beyond the confines of your publisher's offices, make sure you send inscribed copies to your important collaborators. In the publisher's offices (where they already have the book, after all), make a good gesture. Whether you just send thank-you cards (real cards, with real handwritten notes, not just cheesy email) to the relevant people, or splurge on bottles of wine, flowers, cookies, or whatever else is most appropriate (some flies you've tied are a great gift to an editor who fishes), make sure you reinforce your expressions of gratitude this one more time.

Last, now that you're an author, celebrate the occasion. Finishing the manuscript, signing a contract, and especially receiving that first copy of the published book—these are all milestones, worthy of an appropriate beverage or two, or a nice dinner with your spouse or partner. The appearance of the published book you might want to celebrate a little more broadly; invite some friends, thinking especially of those who helped you, over for a little party. Don't be stingy; have a box of copies of your book on hand to inscribe and give to these people who keep you going.

Last, there's a good chance that writing your book got you thinking of others you might write. Whether or not you have the next book in mind, give yourself a few days to catch your breath, and then start thinking about where you'll go from here.

The world can never have too many books. Make sure some of them are yours.

Acknowledgments

As always, first we thank our respective spouses, Joan Raymond and Marsha Karle, for their unfailing encouragement and patience, and for sticking with us all these years.

We would be remiss in a book of this nature were we not to acknowledge the editors at almost countless magazines and book publishers, from whom we've learned so much about the arts, crafts, and general trade of writing during our combined century-plus of writing and publishing.

James Thull, special collections librarian at Montana State University's Trout and Salmonid Collection, Bozeman, and Kirsti Scutt Edwards, collections manager at the American Museum of Fly Fishing, Manchester, Vermont, welcomed and facilitated our use of those two extraordinary libraries for researching information about specific titles discussed in the book and other publishing matters.

Writer-editor Lisa Reuter, Bisbee, Arizona, brought us up to date on the evolving status of contract editing in the rapidly changing realm of self-publishing. Michael S. Gross, director of legal services at the Authors Guild, Inc., provided essential legal advice relating to publishing contracts.

Randy Raymond solved several apparently insoluble computer problems and helped us in countless other ways.

We would also be remiss if we did not gratefully credit our artist friend Al Hassall for the illustration on the book's cover and the chapter heading illustration used throughout the book.

Finally, we owe special thanks to Judith Schnell and the whole crew at Lyons for taking on such an unusual—not to say unprecedented—book as this. Even after writing and publishing so many books, we have never lost our deep appreciation for the people who make the next one possible.

Bibliography

This is a list only of books we mention by title in the text, rather than a list of recommended reading or a list of the "best" fishing books (whatever that means). We list here the original editions except in the cases of (1) some books that we specifically cite or discuss a later edition, and (2) modern general (i.e., non-fishing) reference books for which we have tried to list the most preferable, or at least the most recent, updated edition.

Anonymous (mistakenly attributed to a Barnes or Berners). *A Treatise of Fishing with an Angle* (spelling modernized). First published as part of the second *Book of St. Albans*. London, Wynkyn de Worde, 1496 (a first *Book of St. Albans*, published in 1486, primarily about other sporting subjects, did not include the fishing treatise). The first versions of the fishing treatise, such as this one, were published with a heading that read, "Here beginneth a treatise of fishing with an angle" as the only title. Later editions, when the fishing treatise was published separately, reduced this title to *The Treatise of Fishing with an Angle*, which is how we generally cite it today.

Associated Press. *The Associated Press Stylebook: 2022–2024*. New York: Basic Books, 2022. As with the other style books, look for the most recently updated edition.

Atherton, Maxine. *The Fly Fisher and the River*. New York: Skyhorse, 2016.

Bainbridge, George. *The Fly-fisher's Guide*. Liverpool: Printed for the author by G.F. Harris's Widow and Brothers, 1816.

Bates, Joseph D. *Streamers and Bucktails*. New York: Alfred Knopf, 1980.

Beasley, Beau. *Fly Fishing the Mid-Atlantic*. Tucson: No Nonsense Fly Fishing Guidebooks, 2011.

Beelart, Joseph, Jr. *Howells: The Bamboo Fly Rods & Fly-Fishing Legacy of Gary H. Howells*, Cincinnati: The Whitefish Press, 2020.

Bergman, Ray. *Trout*. Philadelphia: Penn Publishing, 1938.

Berryman, Jack W. *Fly Fishing Pioneers and Legends of the Northwest*. Seattle: Northwest Fly Fishing, LLC, 2006

Block, Erin. *By a Thread: A Retrospective on Women and Fly Tying*. Cincinnati: The Whitefish Press, 2016.

———. *The View from Coal Creek: Reflections on Fly Rods, Canyons and Bamboo*. Cincinnati: The Whitefish Press, 2012.

Born, Steve, Bill Songnozi, and Jeff Mayers. *Exploring Wisconsin Trout Streams: The Angler's Guide*. Madison: University of Wisconsin Press, 1997.

Bowlker, Richard. *The Art of Angling Improved in All Its Parts. Especially Fly Fishing*. Worcester, UK: M. Olivers, circa 1746. Charles Bowlker, the son, is sometimes credited with various early and later editions as either author or coauthor.

Brewer, Robert. *Writer's Market 100th Edition*. New York: Writer's Digest Books, 2021. As with other reference works, look for the most recent edition. This one is updated annually.

Brookes, Richard. *The Art of Angling, Rock and Sea Fishing*. London: John Watts, 1740.

Brooks, Charles. *Nymph Fishing for Larger Trout*. New York: Crown, 1976.

Brown, Darren. *Field Guide to Fishing Knots*. Belgrade, MT: Wilderness Adventures Press, 2003.

Browne, Jen. *Trout Culture: How Fly Fishing Forever Changed the Rocky Mountain West*. Seattle: University of Washington Press, 2015.

Browne, Topher. *Atlantic Salmon Magic*. Bothell, WA: Wild River Press 2011.

Browning, Mark. *Haunted by Waters: Fly Fishing in North American Literature*. Athens: Ohio University Press, 1998.

Burton, Mallory. *Reading the Water: Stories and Essays of Fly Fishing and Life*. Sandpoint, ID: Keokee Books, 1995.

———. *Green River Virgins and Other Passionate Anglers*. New York: Lyons Press, 2000.

Cairncross, Martin, and John Dawson *Trout Fly Fishing: An Expert Approach.* Lanham, MD: Derrydale Press, 2001.
Cannon, David. *Fly Fishing Georgia.* Tucson: No Nonsense Fly Fishing Guidebooks, 2009.
Chatham, Russell. *The Angler's Coast.* New York: Doubleday, 1976.
———, ed. *Silent Seasons.* Santa Monica, CA: Waverly Books, 1978.
Chocklett. Blane. *Game Changer: Tying Flies that Look and Swim Like the Real Thing,* Boiling Springs, PA: Headwater Books, 2020.
Coggins, David. *The Optimist: A Case for the Fly-Fishing Life.* New York: Scribner, 2021.
Combs, Trey. *Steelhead Fly Fishing.* New York: Lyons & Burford, 1991.
Cowgill, Kent. *Sunlit Riffles and Shadowed Runs: Stories of Fly Fishing in America.* Madison: University of Wisconsin Press, 2012.
Craig, Martin, ed., *Fly Fishing in Northern New Mexico.* Albuquerque: University of New Mexico Press, 2002.
Cramer, Jim. *Become a Thinking Fly Tier.* Tucson: No Nonsense Fly Fishing Guidebooks, 2013.
Dauble, Dennis. *Chasing Ghost Trout.* Sandpoint, ID: Keokee Books, 2021.
Dawson, George. *The Pleasures of Angling with Rod and Reel for Trout and Salmon.* New York: Sheldon, 1876.
DeMott, Robert. *Astream: American Writers on Fly Fishing.* New York: Skyhorse, 2014.
———. *Angling Days: A Fly Fisher's Journals.* New York: Skyhorse, 2016.
Fersen, Paul. *Great Fishing Lodges of North America: Fly Fishing's Finest Destinations.* New York: Rizzoli International Publications, 2010.
Flick, Art. *Art Flick's Streamside Guide to Naturals and Their Imitations.* New York: Putnam, 1947.
Francis, Austin. *Catskill Rivers.* New York: Nick Lyons Books, 1983.
Francis, Francis. *A Book on Angling.* London: Longmans, 1867.
Franke, Floyd. *Fish on! A Guide to Playing and Landing Big Fish on a Fly.* Lanham, MD: Derrydale Press, 2003.
Gay, John. *Poems on Several Occasions.* London: Tonson, 1720.
Gierach, John. *Trout Bum.* Boulder:Pruett Publishing, 1986.
Gingrich, Arnold. *The Well-Tempered Angler.* New York: Alfred A. Knopf, 1965.

———, ed. *American Trout Fishing*. New York: Alfred A. Knopf, 1966.

———. *The Fishing in Print: A Guided Tour through Five Centuries of Angling Literature*. New York: Winchester Press, 1974.

Girard, Jerry. *Casting About: The Complete Castabout Columns on Fly Rods and Fly-Fishing History*. Cincinnati:T he Whitefish Press, 2022.

Gordon, Sid. *How to Fish from Top to Bottom*. Mechanicsburg, PA: Stackpole Books, 1955.

Greene, Harry Plunkett. *Where the Bright Waters Meet*. London: Philip Allan, 1924.

Grove, Alvin R. ("Bus"). *The Lure and Lore of Trout Fishing*. Harrisburg, PA: Stackpole Books, 1951.

Grover, Quinn. *Wilderness of Hope: Fly Fishing and Public Lands in the American West*. Lincoln: University of Nebraska Press, 2019.

Haig-Brown, Roderick. *The Western Angler: An Account of Pacific Salmon and Western Trout*. New York: Derrydale Press, 1939.

———. *Return to the River*. New York: William Morrow, 1941.

———. *A River Never Sleeps*. New York: William Morrow, 1946.

———. *Measure of the Year*. New York: William Morrow, 1950.

———. *Fisherman's Spring*. New York: William Morrow, 1951.

———. *Fisherman's Winter*. New York: William Morrow, 1954.

———. *Fisherman's Summer*. New York: William Morrow, 1959.

———. *Fisherman's Fall*. New York, William Morrow, 1964.

———, ed. by Valerie Haig-Brown. *To Know a River: A Haig-Brown Reader*. New York: Lyons & Burford, 1996.

———. *The Seasons of a Fisherman*. New York: Lyons Press, 2000.

Haig-Brown, Valerie Deep *Currents: Roderick and Ann Haig-Brown*, Victoria, BC: Orca Book Publishers, 1998).

Halverson, Anders. *An Entirely Synthetic Fish*. New Haven, CT: Yale University Press, 2010.

Hanley, Ken. *Fly Fishing California*. Tucson: No Nonsense Fly Fishing Guidebooks, 2007.

Herd, Andrew. *The History of Fly Fishing*. Ellesmere, UK: The Medlar Press Ltd., 2019.

Herter, George Leonard. *Professional Fly Tying, Spinning and Tackle Making Manual and Manufacturer's Guide*. Waseca, WI: the author, 1941.

Hilyard, Grayling. *Carrie Stevens: Maker of Rangeley Favorite Trout and Salmon Flies*. Mechanicsburg, PA: Stackpole Books, 2000.

Hogan, Austin, and Paul Schullery. *The Orvis Story*. Manchester, VT: the Orvis Company, 1981.

Hudson, Steve. *Chattahoochee Trout: The Definitive Guide to Chattahoochee Trout Fishing*. Roswell, GA: Chattahoochee Media Group, 2017.

Hudson, Steve. *Fly Fishing Georgia's Toccoa River,* Roswell, GA: Chattahoochee Media Group, 2015.

Hughes, Henry. *The Art of Angling: Poems about Fishing*. New York: Alfred A. Knopf, 2011.

Humphrey, William. *Home from the Hill*. New York: Alfred A. Knopf, 1958.

———. *My Moby Dick*. New York: Doubleday, 1978.

Jennings, Preston. *A Book of Trout Flies*. New York: Derrydale Press, 1935.

Johnson, Kirk Wallace. *The Feather Thief: Beauty, Obsession and the Natural History Heist of the Century*. New York: Viking, 2018.

Johnson, Rex, and Ronald Smorynski. *Fly Fishing in Southern New Mexico*. Albuquerque: University of New Mexico Press, 1998.

Jordan, Will. *Flyfishers Guide to Arizona*. Belgrade, MT: Wilderness Adventures Press, 2008.

Kaminsky, Peter. *The Fly Fisherman's Guide to the Meaning of Life*. New York: Skyhorse, 2008.

Kantner, Steve. *Fifty Women Who Fish*. Bothell, WA: Wild River Press, 2020.

Karas, Nick. *Brook Trout: A Thorough Look at North America's Great Native Trout*. New York: Lyons Press, 1997.

Keane, Martin. *Classic Rods and Rodmakers*. Clinton, NJ: Winchester Press, 1976.

Kelly, Mary Kefover. *The Origins of American Angling: Essays on the History of American Fishing and Tackle*. Cincinnati: The Whitefish Press, 2007.

Koch, Ed. *Fishing the Midge*. New York: Freshet Press, 1972.

Kreh, Lefty. *Fly Casting with Lefty Kreh*. Philadelphia: Lippincott, 1974.

———. *Fly Fishing in Salt Water*. New York: Nick Lyons Books, 1986.

———. *My Life Was This Big*. New York: Skyhorse, 2014.

Kreiger, Mel. *The Essence of Fly Casting*. San Francisco: Club Pacific, 1987.

Kroll, Charles. *Squaretail*. New York: Vantage Press, 1972.

Kurlansky, Mark. *Salmon: A Fish, the Earth and the History of Their Common Fate*. Ventura, CA: Patagonia Books, 2020.

LaFontaine, Gary. *Caddisflies*. New York: Nick Lyons Books, 1981.

Leeson, Ted. *The Habit of Rivers*. New York: Lyons & Burford, 1994.

Leiser, Eric, C. Boyd Pfeiffer and Jack Gartside. *The Complete Book of Fly Tying*. New York: Skyhorse, 2014.

Leiser, Eric. *The Dettes: A Catskill Legend*. Fishkill, NY: Willowkill, 1992.

Lilly, Bud, and Paul Schullery. *A Trout's Best Friend*. Boulder: Pruett Publishing, 1985.

———. *Bud Lilly's Guide to Fly Fishing the New West*. Portland, OR: Frank Amato, 2000.

Lyons, Nick, ed. *Fisherman's Bounty*. New York: Crown, 1970.

———. *Fishing Widows*. New York: Crown, 1974.

———, ed., *The Gigantic Book of Fishing Stories*. New York: Skyhorse, 2007.

Lyons, Nick, and Mari Lyons. *A Fly Fisher's World*. New York: Skyhorse, 2013.

Maclean, Norman. *A River Runs Through It and Other Stories*. Chicago: University of Chicago Press, 1976.

Marinaro, Vincent. *A Modern Dry-Fly Code*. New York: Putnam's, 1950.

Maxwell, Jessica. *I Don't Know Why I Swallowed the Fly*. New York: Lyons Press, 1998.

Mayers, Jeff, and Tom Wendelberg. *Catching Big Fish on Light Fly Tackle*. Madison: University of Wisconsin Press, 2001.

McClane, A. J. *The Practical Fly Fisherman*. Englewood Cliffs, NJ: Prentice Hall, 1953.

———, ed., *McClane's Standard Fishing Encyclopedia*. New York: Holt, Rinehart & Winston, 1965.

McDonald, John, ed., *The Compleat Fly Fisherman: The Notes and Letters of Theodore Gordon*. New York: Charles Scribner's Sons, 1947.

———. *Quill Gordon*. New York: Alfred A. Knopf, 1972.

———. *The Origins of Angling*. Garden City, NY: Doubleday, 1963.

McGuane, Thomas. *The Longest Silence*. New York: Alfred A. Knopf, 2000.
Merwin, John. *The Battenkill*. New York: Lyons & Burford, 1993.
———. *The New American Trout Fishing*. New York: Macmillan, 1997.
Meyers, Steven J. *Notes from the San Juan*. New York: Lyons & Burford, 2014.
Middleton, Harry. *The Earth Is Enough: Growing Up in a World of Trout and Old Men*. New York: Simon & Schuster, 1989.
———. *Rivers of Memory*. Boulder: Pruett Publishing, 1993.
Miller, Alfred W. (pseudonym Sparse Grey Hackle). *Fishless Days, Angling Nights* Essex, CT: Lyons Press, 1971.
Montgomery, M. R. *Many Rivers to Cross*. New York: Simon & Schuster, 1995.
Morris, Holly. *Uncommon Waters: Women Write about Fishing*. New York: Seal Press, 1991.
———. *A Different Angle: Fly Fishing Stories by Women*. New York: Seal Press, 1995.
Nemes, Sylvester. *The Soft-Hackled Fly*. Greenwich, CT: Old Chatham Press, 1975.
Norris, Thaddeus. *The American Angler's Book*. Philadelphia: E.H. Butler/Sampson Low, 1864.
O'Brien, Geoffrey, and John Bartlett. *Bartlett's Familiar Quotations*. Boston: Little, Brown and Company, 1922.
Olch, Jonathan. *A Passion for Permit* (two volumes). Bothell, WA: Wild River Press, 2016.
Page, Margot. *Little Rivers*. New York: Lyons & Burford, 1995.
Perkins, Leigh. *A Sportsman's Life, How I Built Orvis by Mixing Business and Sport.* New York: Lyons Press, 2022.
Piper, Ti. *Fishing in New Mexico*. Albuquerque: University of New Mexico Press, 1989.
Poland, Tim. *Yellow Stonefly*. Athens: Ohio University/Swallow Press, 2018.
Proper, Datus. *What the Trout Said*. New York: Alfred A. Knopf, 1982.
Quinnett, Paul. *Pavlov's Trout*. Sandpoint, ID: Keokee Books, 1994.
Ramsay, Henry, Mike Heck and Len Lichvar. *Keystone Fly Fishing: The Ultimate Guide to Pennsylvania's Best Water*. Boiling Springs, PA: Headwater Books, 2017.

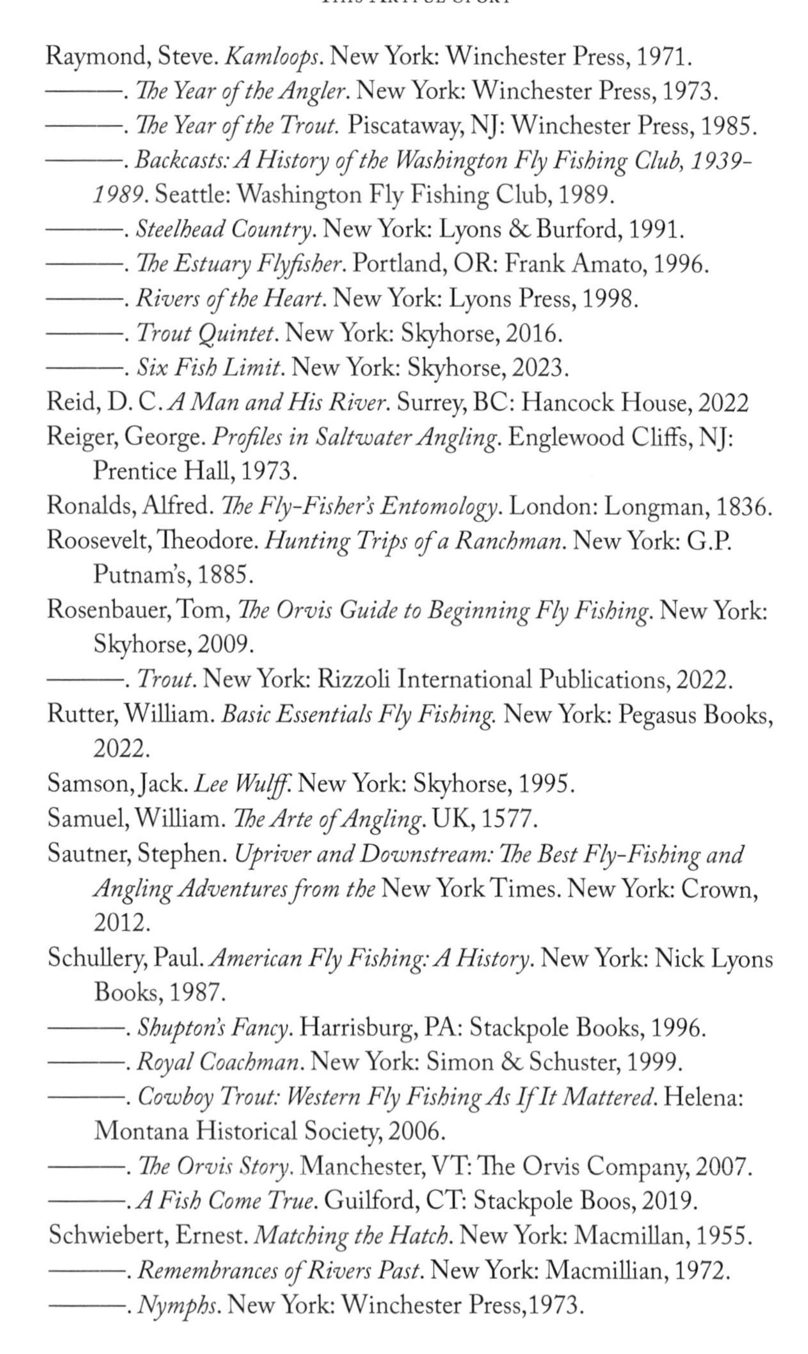

Raymond, Steve. *Kamloops*. New York: Winchester Press, 1971.

———. *The Year of the Angler*. New York: Winchester Press, 1973.

———. *The Year of the Trout.* Piscataway, NJ: Winchester Press, 1985.

———. *Backcasts: A History of the Washington Fly Fishing Club, 1939-1989*. Seattle: Washington Fly Fishing Club, 1989.

———. *Steelhead Country*. New York: Lyons & Burford, 1991.

———. *The Estuary Flyfisher*. Portland, OR: Frank Amato, 1996.

———. *Rivers of the Heart*. New York: Lyons Press, 1998.

———. *Trout Quintet*. New York: Skyhorse, 2016.

———. *Six Fish Limit*. New York: Skyhorse, 2023.

Reid, D. C. *A Man and His River*. Surrey, BC: Hancock House, 2022

Reiger, George. *Profiles in Saltwater Angling*. Englewood Cliffs, NJ: Prentice Hall, 1973.

Ronalds, Alfred. *The Fly-Fisher's Entomology*. London: Longman, 1836.

Roosevelt, Theodore. *Hunting Trips of a Ranchman*. New York: G.P. Putnam's, 1885.

Rosenbauer, Tom, *The Orvis Guide to Beginning Fly Fishing*. New York: Skyhorse, 2009.

———. *Trout*. New York: Rizzoli International Publications, 2022.

Rutter, William. *Basic Essentials Fly Fishing*. New York: Pegasus Books, 2022.

Samson, Jack. *Lee Wulff*. New York: Skyhorse, 1995.

Samuel, William. *The Arte of Angling*. UK, 1577.

Sautner, Stephen. *Upriver and Downstream: The Best Fly-Fishing and Angling Adventures from the* New York Times. New York: Crown, 2012.

Schullery, Paul. *American Fly Fishing: A History*. New York: Nick Lyons Books, 1987.

———. *Shupton's Fancy*. Harrisburg, PA: Stackpole Books, 1996.

———. *Royal Coachman*. New York: Simon & Schuster, 1999.

———. *Cowboy Trout: Western Fly Fishing As If It Mattered*. Helena: Montana Historical Society, 2006.

———. *The Orvis Story*. Manchester, VT: The Orvis Company, 2007.

———. *A Fish Come True*. Guilford, CT: Stackpole Boos, 2019.

Schwiebert, Ernest. *Matching the Hatch*. New York: Macmillan, 1955.

———. *Remembrances of Rivers Past*. New York: Macmillian, 1972.

———. *Nymphs*. New York: Winchester Press,1973.

———. *Death of a Riverkeeper*. New York: Dutton, 1980.
———. *A River for Christmas*. Lexington, MA: Stephen Greene, 1988.
———. *The Henryville Flyfishers: A Chronicle of American Fly Fishing*. Far Hills, NJ: Meadow Run, 1998.
Scotcher, George. *The Fly Fisher's Legacy*. Chepstow, UK: M. Willet, circa 1810.
Shepard, Odell. *Thy Rod and Thy Creel*. Hartford, CT: Mitchell, 1930.
Siegal, Allan M., and William Connolly. *The New York Times Manual of Style and Usage*. New York: Crown, 2015. As with all writing guides and similar reference books, find the most recently updated edition.
Skues, G. E. M. *The Way of a Trout with a Fly*. London: Black, 1921.
Snyder, Samuel, Bryon Borgelt, and Elizabeth Tobey, eds., *Backcasts: A Global History of Fly Fishing and Conservation*. Chicago: University of Chicago Press, 2016.
Steinberg, Michael K. *Searching for Home Waters: A Brook Trout Pilgrimage,* Athens: University of Georgia Press, 2023.
Stoner, Lew, and Ted Trueblood. *The Lew Stoner-Ted Trueblood Letters*, Vol. 2. Cincinnati: The Whitefish Press, 2021.
Strunk, William, Jr., and E. B. White. *The Elements of Style*. Fourth edition. New York: Pearson Education, Inc./Macmillan, 1999 (sometimes cited as 2000), foreword by Roger Angell. Be careful to get this edition. The book exists in variants almost beyond counting, including three excellent previous "official" editions (1959, 1972, 1979) that were overseen and updated by White, who died in 1985. The fourth edition, as cited above, contains all of Strunk and White's advice but is also attentive to some modern challenges—such as changing ideas of gender terminology. Beware numerous other so-called editions, especially now that the very earliest versions of the book, published when only Strunk was the author, have fallen into the public domain. Quick and dirty unauthorized reprints of those earliest versions, often containing "updates" and other anonymous revisions, abound from online sellers.
Supinsky, Matt. *Selectivity: The Theory and Method of Fly Fishing for Fussy Trout, Salmon and Steelhead.* Boiling Springs, PA: Headwater Books, 2014.
Swisher, Doug, and Carl Richards. *Selective Trout*. New York: Crown, 1971.

Tallack, Mallachy. *Illuminated by Water: Fly Fishing and the Allure of the Natural World*. New York: Pegasus Books. 2022.

Talleur, Richard W. *Mastering the Art of Fly Tying*. Harrisburg, PA: Stackpole Books, 1979.

———. *The Fly Tyer's Primer*. New York: Lyons & Burford, 1986.

———. *Modern Fly-Tying Materials*. New York: Lyons & Burford, 1995.

Tapply, William G. *A Fly-Fishing Life*. New York: The Lyons Press, 1997.

———. *Pocket Water*. New York: The Lyons Press, 2001.

———. *Every Day Was Special*. New York: Skyhorse, 2014.

Tomine, Dylan. *Headwaters: The Adventures, Obsessions and Evolution of a Fly Fisherman*. Ventura, CA: Patagonia Books, 2022.

Traver, Robert (pen name of John Voelker). *Anatomy of Murde*r. New York: St. Martin's Press, 1958.

———. *Trout Madness*. New York: St. Martin's Press, 1960.

———. *Trout Magic*. New York: Crown, 1974.

University of Chicago Press Editorial Staff. *The Chicago Manual of Style, 17th Edition*. Chicago: University of Chicago Press, 2017. As with all writing guides and similar reference books, find the most recently updated edition.

Van Dyke, Henry. *Little Rivers*. New York: Scribner's, 1885.

Van Put, Ed. *The Beaverkill: The History of a River and Its People*. Essex, CT: The Lyons Press, 1996.

———. *Trout Fishing in the Catskills*. New York: Skyhorse, 2014.

Venables, Col. Robert. *The Experienced Angler: or angling improved*. London: Richard Marriot, 1662.

Veniard, John. *Fly-Dressing Materials*. Salisbury, UK: J&J Head, 1977.

Wakefield, Jacqueline. *Flytying Techniques*. London: A.&C. Black, 1980.

Walden, Howard, II. *Upstream and Down*. New York: Derrydale Press, 1938.

Warner, Charles Dudley, *In the Wilderness*. Boston: Houghton, Osgood, 1878.

Weamer, Paul. *The Bug Book: A Fly Fisher's Guide to Trout Stream Insects*. Boiling Springs, PA: Headwater Books, 2016.

Westwood, Thomas, and Thomas Satchell. *Bibliotheca Piscatoria, A Catalogue of Books on Angling, the Fisheries, and Fish-Culture*. London: W. Satchell, 1883.

Wetherell, W. D., *Vermont River*. New York: Nick Lyons Books, 1984.

———. *Upland Stream*. Boston: Little, Brown and Company, 1991.

———. *One River More*. New York: Nick Lyons Books, 1998.

Wethern, Robert. ed. *The Creel: North Umpqua Edition*. Milwaukie, OR: Frank Amato Publications, Inc., 2008.

Wetzel, Charles M. *Trout Flies, Naturals and Imitations*. Mechanicsburg, PA: Stackpole Books,1955.

Whitlock, Dave. *Guide to Aquatic Trout Foods*. New York: Nick Lyons Books 1982.

Wickstrom, Gordon. *Notes from an Old Fly Book*. Boulder: University Press of Colorado, 2001.

———. *Late in an Angler's Life*. Albuquerque: University of New Mexico Press, 2004.

Wilderness Adventures Press. *Washington's Best Waters, a Wilderness Adventures Press Map Book*. Belgrade, MT: Wilderness Adventures Press, 2007.

Wright, Leonard M., Jr. *Flutter, Skitter, and Skim: Using the Living Insect as a Guide for Successful Fly Fishing*. Lanham, MD: Derrydale Press, 2001.

Wulff, Joan. *Fly Fishing: Expert Advice from a Woman's Perspective*. Mechanicsburg, PA: Stackpole Books, 1991.

———. *Fly Casting Accuracy*. New York: Lyons Press, 1997.

———. *New Fly Casting Techniques*. New York: Lyons Press, 2012.

Wulff, Lee. *Lee Wulff on Flies*. Harrisburg, PA: Stackpole Books, 1980.

———. *Trout on a Fly*. New York: Nick Lyons Books, 1986.

Zellere, Bob. *Business Traveler's Guide to Fly Fishing in the Western States*. Tucson: No Nonsense Fly Fishing Guidebooks, 1999.

Zinsser, *On Writing Well*. New York: Harper Perennial, 2012.

About the Authors

Paul Schullery learned to fly fish more than 50 years ago while a ranger-naturalist in Yellowstone Park. In 1977 he became the first executive director of The American Museum of Fly Fishing and editor of *The American Fly Fisher*. Later, back in Yellowstone, he was founding editor of *Yellowstone Science*, the park's research journal. A Life Member of Fly Fishers International and Trout Unlimited, Paul is the author of nine fly-fishing books, including *American Fly Fishing: A History*, *The Fishing Life*, and *The Rise*, as well as many other books on nature and conservation. A recipient of the Roderick Haig-Brown Award, Paul was inducted into the Fly Fishing Hall of Fame in 2014. He lives in southern Vermont and is married to the artist Marsha Karle, with whom he has collaborated as author and artist on several books.

Steve Raymond, a native of Bellingham, Washington, had a 30-year career as a reporter, editor and manager at the *Seattle Times*. He also edited two magazines, *The Flyfisher* and *Fly Fishing in Salt Waters*, and reviewed fishing books for several publications. A charter and honorary life member of the Federation of Fly Fishers (now called Fly Fishers International), he is author of a dozen fly-fishing books, including two award-winning titles, *The Year of the Angler* and *The Year of the Trout*. He received the prestigious Roderick Haig-Brown Award for significant contributions to angling literature, and his work has appeared in nine anthologies and at least 24 magazines. He was inducted into the Fly Fishing Hall of Fame in October 2022. Raymond and his wife, Joan, reside on an old farm on Whidbey Island in northern Puget Sound.